Computational Modeling of Materials and Processing

Related titles published by the American Ceramic Society:

Ceramic Manufacturing Practices and Technologies (Ceramic Transactions Volume 70)
Edited by Basavaraj Hiremath, Tapan Gupta, and K.M. Nair
© 1997, ISBN 1-57498-024-6

Ceramic Joining (Ceramic Transactions Volume 77)
Edited by Ivar E. Reimanis, Charles H. Henager, and Antoni P. Tomsia
© 1997, ISBN 1-57498-022-X

Environmental Issues and Waste Management Technologies in the Ceramic and Nuclear Industries II (Ceramic Transactions Volume 72)
Edited by Vijay Jain and David Peeler
© 1996, ISBN 1-57498-023-8

Advances in Ceramic-Matrix Composites III (Ceramic Transactions Volume 74)
Edited by Narottam P. Bansal and J.P. Singh
© 1996, ISBN 1-57498-020-3

Corrosion of Materials by Molten Glass (Ceramic Transactions Volume 78)
Edited by George Pecoraro, James C. Marra, and John T. Wenzel
© 1996, ISBN 1-57498-019-X

Mass and Charge Transport in Ceramics (Ceramic Transactions Volume 71)
Edited by Kunihito Koumoto, H. Matsubara, and Laurel Sheppard
© 1996, ISBN 1-57498-018-1

Handbook on Continuous Fiber–Reinforced Ceramic Matrix Composites
Edited by Richard L. Lehman, Said K. El-Rahaiby, and John B. Wachtman Jr.
© 1995, ISBN 0-931682-43-6

Handbook on Discontinuously Reinforced Ceramic Matrix Composites
Edited by Keith J. Bowman, Said K. El-Rahaiby, and John B. Wachtman Jr.
© 1995, ISBN 0-931682-42-8

Databook on Mechanical and Thermophysical Properties of Whisker-Reinforced Ceramic Matrix Composites
Christian X. Campbell and Said K. El-Rahaiby
© 1995, ISBN 0-931682-56-8

Databook on Mechanical and Thermophysical Properties of Particulate-Reinforced Ceramic Matrix Composites
Christian X. Campbell and Said K. El-Rahaiby
© 1995, ISBN 0-931682-57-6

Advanced Synthesis and Processing of Composites and Advanced Ceramics II (Ceramic Transactions Volume 79)
Edited by Kathryn V. Logan, Zuhair Munir, and Richard M. Spriggs
© 1996, ISBN 1-57498-17-3

Hybrid Microelectronic Materials (Ceramics Transactions Volume 68)
Edited by K.M. Nair and V.N. Shukla
© 1996, ISBN 1-57498-013-0

For information on ordering titles published by the American Ceramic Society, or to request a publications catalog, please contact our Customer Service Department at 614-794-5890 (phone), 614-794-5854 (fax), <customersrvc@acers.org> (e-mail), or write to Customer Service Department, 735 Ceramic Place, Westerville, OH 43081. Visit our on-line book catalog and web site at <www.acers.org>.

Role of Ceramics in Advanced Electrochemical Systems (Ceramic Transactions Volume 65)
Edited by Prashant N. Kumta, Gregory S. Rohrer, and U. Balachandran
© 1996, ISBN 1-57498-009-2

Phase Equilibria Diagrams, Phase Diagrams for Ceramists, Volume XI: Oxides
Edited by Robert S. Roth
© 1995, ISBN 0-944904-90-4

Environmental Issues and Waste Management Technologies in the Ceramic and Nuclear Industries (Ceramic Transactions Volume 61)
Edited by Vijay Jain and Ron Palmer
© 1995, ISBN 1-57498-004-1

Crystal Growth of Novel Electronic Materials (Ceramic Transactions Volume 60)
Edited by R.K. Pandey and Ruyan Guo
© 1995, ISBN 1-57498-003-3

Microwaves: Theory and Application in Materials Processing III (Ceramic Transactions Volume 59)
Edited by David E. Clark, Diane C. Folz, Steven J. Oda, and Richard Silberglitt
© 1995, ISBN 1-57498-002-5

High-Temperature Ceramic-Matrix Composites II: Manufacturing and Materials Development (Ceramic Transactions Volume 58)
Edited by A.G. Evans and R. Naslain
© 1995, ISBN 0-944904-99-8

High-Temperature Ceramic-Matrix Composites I: Design, Durability, and Performance (Ceramic Transactions Volume 57)
Edited by A.G. Evans and R. Naslain
© 1995, ISBN 0-944904-98-X

Advanced Synthesis and Processing of Composites and Advanced Ceramics (Ceramic Transactions Volume 56)
Edited by Kathryn V. Logan
© 1995, ISBN 1-57498-000-9

Sol-Gel Science and Technology (Ceramic Transactions Volume 55)
Edited by Edward J.A. Pope, Sumio Sakka, and Lisa Klein
© 1995, ISBN 0-944904-97-1

Science, Technology, and Applications of Colloidal Suspensions (Ceramic Transactions Volume 54)
Edited by James H. Adair, Jon A. Casey, Clive A. Randall, and Sridhar Venigalla
© 1995, ISBN 0-944904-96-3

Materials and Processes for Wireless Communications (Ceramic Transactions Volume 53)
Edited by T. Negas and H. Ling
© 1995, ISBN 0-944904-94-7

Low-Expansion Materials (Ceramic Transactions Volume 52)
Edited by David P. Stinton and Santosh Y. Limaye
© 1995, ISBN 0-944904-92-0

Ceramic Processing Science and Technology (Ceramic Transactions Volume 51)
Edited by Hans Hausner, Gary L. Messing, and Shin-ichi Hirano
© 1995, ISBN 0-944904-89-0

Design for Manufacturability of Ceramic Components (Ceramic Transactions Volume 50)
Edited by Asish Ghosh, Basavaraj Hiremath, and John Halloran
© 1995, ISBN 0-944904-88-2

Manufacture of Ceramic Components (Ceramic Transactions Volume 49)
Edited by Basavaraj Hiremath, Allan Bruce, and Asish Ghosh
© 1995, ISBN 0-944904-87-4

Bioceramics: Materials and Applications (Ceramic Transactions Volume 48)
Edited by Gary Fischman, Alexis Clare, and Larry Hench
© 1995, ISBN 0-944904-82-3

Magnetic Ceramics (Ceramic Transactions Volume 47)
Edited by Bhaskar B. Ghate and John J. Simmins
© 1995, ISBN 0-944904-81-5

Advances in Ceramic-Matrix Composites II (Ceramic Transactions Volume 46)
Edited by J.P. Singh and Narottam P. Bansal
© 1994, ISBN 0-944904-80-7

Environmental and Waste Management Issues in the Ceramic Industry II (Ceramic Transactions Volume 45)
Edited by Dennis Bickford, Steven Bates, Vijay Jain, and Gary Smith
© 1994, ISBN 0-944904-79-3

Materials Processing and Design: Grain-Boundary-Controlled Properties of Fine Ceramics II (Ceramic Transactions Volume 44)
Edited by Koichi Niihara, Kozo Ishizaki, and Mitsuo Isotani
© 1994, ISBN 0-944904-78-5

Ferroic Materials: Design, Preparation, and Characteristics (Ceramic Transactions Volume 43)
Edited by A.S. Bhalla, K.M. Nair, I.K. Lloyd, H. Yanagida, and D.A. Payne
© 1994, ISBN 0-944904-77-7

Silicon-Based Structural Ceramics (Ceramic Transactions Volume 42)
Edited by Brian W. Sheldon and Stephen C. Danforth
© 1994, ISBN 0-944904-76-9

Grain Boundaries and Interfacial Phenomena in Electronic Ceramics (Ceramic Transactions, Volume 41)
Edited by Lionel M. Levinson and Shin-ichi Hirano
© 1994, ISBN 0-944904-73-4

Cement Technology (Ceramic Transactions Volume 40)
Edited by E.M. Gartner and H. Uchikawa
© 1994, ISBN 0-944904-72-6

Advances in Ceramic-Matrix Composites (Ceramic Transactions Volume 38)
Edited by Narottam P. Bansal
© 1993, ISBN 0-944904-69-6

Microwaves: Theory and Application in Materials Processing II (Ceramic Transactions Volume 36)
Edited by David E. Clark, Wayne R. Tinga, and Joseph R. Laia Jr.
© 1993, ISBN 0-944904-66-1

Structural Ceramics Joining II (Ceramic Transactions Volume 35)
Edited by Arthur J. Moorhead, Ronald E. Loehman, and Sylvia M. Johnson
© 1993, ISBN 0-944904-65-3

Functionally Gradient Materials (Ceramic Transactions Volume 34)
Edited by J. Birch Holt, Mitsue Koizumi, Toshio Hirai, and Zuhair A. Munir
© 1993, ISBN 0-944904-64-5

Materials in Microelectronic and Optoelectronic Packaging (Ceramic Transactions, Volume 33)
Edited by Hung C. Ling, Koichi Niwa, and Vishwa N. Shukla
© 1993, ISBN 0-944904-63-7

Dielectric Ceramics: Processing, Properties, and Applications (Ceramic Transactions, Volume 32)
Edited by K.M. Nair, J.P. Guha, and A. Okamoto
© 1993, ISBN 0-944904-60-2

Phase Equilibria Diagrams, Phase Diagrams for Ceramists, Volume IX: Semiconductors and Chalcogenides
Edited by Gerald B. Stringfellow
© 1992, ISBN 0-944904-50-5

Microwaves: Theory and Application in Materials Processing (Ceramic Transactions Volume 21)
Edited by David E. Clark, Frank D. Gac, and Willard H. Sutton
©1991, ISBN 0-944904-43-2

Experimental Techniques of Glass Science
Edited by Catherine J. Simmons and Osama El-Bayoumi
© 1993, ISBN 0-944904-58-0

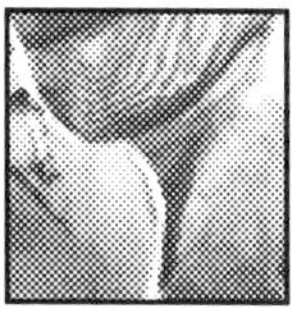

Ceramic Transactions
Volume 69

Computational Modeling of Materials and Processing

Edited by

Joseph H. Simmons
University of Florida

Edwin R. Fuller, Jr.
National Institute of Standards and Technology

Alan L. Dragoo
U.S. Department of Energy

Edward J. Garboczi
National Institute of Standards and Technology

Published by
The American Ceramic Society
735 Ceramic Place
Westerville, Ohio 43081

Proceedings of the Computational Modeling of Materials and Processing symposium, held at the 97th Annual Meeting of the American Ceramic Society in Cincinnati, Ohio, May 1–4, 1995.

For information on ordering titles published by the American Ceramic Society, or to request a publications catalog, please call 614-794-5890.

Printed in the United States of America.

1 2 3 4 5–01 00 99 98 97
ISSN 1042-1122
ISBN 1-57498-039-4

Contents

Structure and Phase Equilibria

Glass Processing

Ceramic Processing

Microstructural Evolution and Grain Growth

Diffusion and Mass Transport

Fracture

Preface

This volume contains selected papers that were presented at the symposium on Computational Modeling of Materials and Processing, which was held as part of the 97th Annual Meeting of The American Ceramic Society from April 30–May 3, 1995, in Cincinnati, Ohio. The symposium was sponsored by the Glass and Optical Materials, Basic Science, Engineering Ceramics, and Cements divisions.

The objective of the symposium was to bring together scientists and engineers from a variety of disciplines to focus on computational aspects of materials and processing. Interactions between theoretical and computational research and experimental research was encouraged, and indeed many of the papers reflect this flavor. Topics ranged from computational modeling at the atomistic level to that at the microstructural level to that at the continuum level. Modeling at the atomistic and molecular level, included calculations of glass and crystalline structure, of surface and interface structure, and of phase equilibria. Topics at the microstructural, or mesoscopic level, included grain growth, microstructural evolution, diffusion, and fracture. At the continuum level, topics included modeling of glass and ceramic processing.

Bringing together such myriad disciplines and size scales into a single symposium was found to be both synergistic and beneficial. The topics at the meeting generated much lively discussion as well as fruitful interactions. The collection of high-quality research papers represents current state-of-the-art in the area of computational materials science and engineering.

Finally, we express our sincere appreciation and thanks to Sarah Godby and the publication staff at The American Ceramic Society for the devoted cooperation and assistance which they provided in preparing this volume.

Joseph H. Simmons
Edwin R. Fuller, Jr.
Alan L. Dragoo
Edward J. Garboczi

Atomistic and Molecular Fundamentals

INTERATOMIC POTENTIALS FOR COVALENT MATERIALS FROM A LOCAL APPROXIMATION TO TIGHT BINDING

A.F. Voter, J.D. Kress, and R.N. Silver
Theoretical Division
Los Alamos National Laboratory
Los Alamos, NM 87545

ABSTRACT

We present a local approximation to total energy tight binding (TB) designed to be suitable as a semiempirical potential form for covalent and metallic-covalent materials. Chebyshev polynomial moments of the Hamiltonian matrix are generated in a stable and efficient manner through recursive matrix-vector multiplies. The electronic density of states is approximated by a Chebyshev polynomial expansion with Gibbs damping which maintains positivity. The computational scaling is made linear in the number of atoms by truncating the moment computation at a certain range about each atom. Energy derivatives for molecular dynamics are obtained from a matrix-polynomial derivative relation. The method converges to exact TB as the number of moments and the truncation range are increased. The convergence properties are tested on silicon.

INTRODUCTION

With the increasing use of atomistic simulations for materials problems, there is a need for new, general forms of interatomic potential, especially for the type of bonding in covalent materials and advanced intermetallics. The total-energy tight-binding (TB) method is increasingly viewed as a possible solution to this problem. While TB is much less computationally demanding than first-principles calculations, it nonetheless suffers from a computational bottleneck in the Hamiltonian diagonalization step, which scales as the number of atoms (or number of basis functions, N) to the third power.

Recently, considerable attention has been given to developing approximations that reduce the scaling of TB from N^3 to N [1-9]. Typical break-even sizes, for which the N-scaling calculation becomes faster than the direct N^3 calculation, range from 50 to a few hundred atoms. Because any approach of this type requires a compromise, a given N-scaling method can have advantages over others, depending on the nature of the application. In the present work, we develop a method that should be suitable as a short-ranged, many-body interatomic potential. It is explicitly local in character, with a deterministic energy and exact energy derivatives. The N-scaling nature is achieved by a local truncation of the environment around each atom. (Infinite truncation range gives an N^2-scaling method [10,11].) The approach is closely related to that presented by Goedecker and Columbo [7]; the main differences are in the definition of the local truncation procedure, and the fact that exact energy derivatives are obtained in the present work.

TOTAL ENERGY TIGHT BINDING METHOD

We first briefly review the orthogonal tight-binding method [12] as it is currently used in atomistic simulations. Given a system of N_{atom} atoms at a specified geometry, one constructs the one-electron Hamiltonian matrix (**H**) over the N-dimensional basis consisting of a valence basis set on each atom. The off-site matrix elements of **H** generally have a simple, two-center form, with a parameterized radial shape for each possible angular momentum combination [e.g., $h_{ss\sigma}(r)$, $h_{sp\sigma}(r)$, $h_{pp\sigma}(r)$, and $h_{pp\pi}(r)$, for an sp^3 basis].

Diagonalizing **H** gives a set of eigenvalues $\{\epsilon_i\}$, which define the electronic density of states (DOS), $n(\epsilon) = \sum_i^N \delta(\epsilon - \epsilon_i)$. The electronic energy is defined as the bond integral over the occupied states of this DOS,

$$E_{elec} = 2\int_{-\infty}^{E_F} \epsilon\, n(\epsilon)\, d\epsilon, \tag{1}$$

where the factor of two accounts for the closed-shell spin state (two electrons per orbital). The Fermi level (E_F) is either prespecified or is determined by requiring that the system have the correct number of valence electrons,

$$N_{val} = 2\int_{-\infty}^{E_F} n(\epsilon)\, d\epsilon. \tag{2}$$

The total energy of the system is then obtained by augmenting E_{elec} with a pair term, $\frac{1}{2}\sum_{i\neq j}\phi(r_{ij})$, where $\phi(r_{ij})$ is a function of the internuclear distance $r_{ij} = |\mathbf{r}_{ij}|$, representing core-core interactions and neglected contributions to the true electronic energy.

LOCAL TRUNCATION

Local subspace truncation, which is responsible for the N-scaling computational dependence, is described and tested in this section. For simplicity, we discuss a system with only one basis function per atom, so that we can refer to basis functions and atoms interchangeably. For each atom i, a local subspace of the full Hamiltonian is defined, containing N_i basis functions. This can be accomplished by retaining only those atoms within a certain physical distance from atom i, a procedure we will refer to as "physical" truncation. Alternatively, one can retain all atoms within a given number (L) of Hamiltonian links from atom i, referred to as "logical" truncation. For example, $L = 1$ truncation retains all atoms within r_{cut} of atom i, where r_{cut} is the cutoff distance for $h(r)$, while $L = 2$ truncation retains all h-linked neighbors of i and all the neighbors of those neighbors.

Within this subspace, we solve the TB electronic structure problem, projecting out the DOS belonging to atom i [$n_i(\epsilon)$]. From the eigenvalues ($\{\epsilon_j\}; j = 1, N_i$) and eigenvectors ($\{C_{jk}\}; j,k = 1, N_i$) obtained from diagonalization of the subspace Hamiltonian, the projected DOS is given by $n_i(\epsilon) = \sum_j^{N_i} C_{ij}^2\delta(\epsilon - \epsilon_j)$.

Note that one can compute the energy of each atom individually, by replacing $n(\epsilon)$ with $n_i(\epsilon)$ in Eq. (1). Also, if desired, one can enforce charge neutrality [via Eq. (2)] atom by atom. A feature of this atom-neutral approach is that it offers a completely local definition of the energy of each atom, as desired for a short-ranged empirical potential.

Because the computational work required to compute each atom DOS depends only on the size of the subspace for that atom, the work required for the whole system scales as N. However, the method as described is not viable for many applications, because calculation of energy derivatives would require differentiation of the eigenvector matrix elements, which cannot be performed analytically. The moment method presented below solves this derivative problem. However, it is useful to first examine the direct-diagonalization approach, because it gives the exact result for a given subspace. We can study the convergence to exact TB as the number of atoms in the subspace (N_i) is increased, without any coupling to the approximations in the moment description.

All calculations presented here employ the silicon tight-binding parameterization of Goodwin, Skinner and Pettifor [13], which has an sp^3 basis set. An abrupt cutoff is imposed between 1st and 2nd neighbor shells for compatibility with previous studies [14,15]. Figure 1 shows the convergence of the electronic energy (1 J = 0.624×10^{19} eV) plotted versus the subspace size, using both the logical truncation (up to L=6) and the physical truncation procedures. Because of the first-nearest-neighbor cutoff in $h(r)$, the first two points (L=1,2) in the logical truncation curve correspond to exactly the same subspaces as for the physical truncation procedure, so the energies coincide. For larger subspaces, the logical truncation continues to converge monotonically toward the 1000-atom periodic supercell value (an approximation to the infinite-system result), passing very close to the energies from 64-atom and 216-atom supercell calculations. In contrast, the physical truncation shows less-satisfactory, nonmonotonic behavior. We infer that logical truncation is preferable to physical truncation, although this assumption should be more thoroughly tested. Mauri and Galli [8] reached a similar conclusion for the truncation of the local orbitals in their N-scaling TB method, although nonmonotonic behavior was not observed.

We have also found rapid convergence of the silicon lattice constant, with an error of less than 0.5% by L=3. The unrelaxed vacancy formation energy shows slower, but reasonable, convergence (errors of -18%, -21%, and -7% at L=3,4 and 5, respectively, relative to the 1000-atom result of $E^f_{vac} = 9.77\times10^{-19}$ J). A previous study using a low-order moment method [15] has shown this to be a difficult property to describe accurately.

KERNEL POLYNOMIAL METHOD

The kernel polynomial method (KPM), implemented with Chebyshev polynomials, offers a controlled, differentiable approximation to the bond integral in (1). This section provides a brief introduction to the KPM, which is presented in detail elsewhere[10,11,16].

Chebyshev polynomials of the first kind, $T_m(x)$, are defined by $T_0(x) = 1, T_1(x) = x$, and the recurrence relation $T_m(x) = 2xT_{m-1}(x) - T_{m-2}(x)$, where the range of x is limited to $-1 \leq x \leq 1$. They obey the orthogonality relation $\int_{-1}^{1} w(x)T_m(x)T_n(x)dx = q_m\delta_{m,n}$, where the weight function is $w(x) = \frac{1}{\sqrt{1-x^2}}$ and $q_m = \pi$ when m=0, and $\pi/2$ otherwise.

We can express an arbitrary function on the interval $\{-1 : 1\}$ as a basis set expansion

$$f(x) = \sum_{m=0}^{\infty} \frac{a_m}{q_m} w(x)T_m(x), \tag{3}$$

where the expansion coefficients are just the Chebyshev moments of the function,

$$a_m = \mu_m = \int_{-1}^{1} f(x)T_m(x)dx. \tag{4}$$

While (3) is exact, we now consider truncating the expansion at M moments to obtain a usable approximation, e.g., to the electronic DOS. After shifting and scaling the energy ($\mathbf{H} = a\bar{\mathbf{H}}+b$) to form $\bar{\mathbf{H}}$, which has an eigenspectrum in the range $\{-1 : 1\}$, we can express the projected DOS for the ith basis function as

$$n_i(\epsilon) \cong \sum_{m=0}^{M} \frac{\mu_{mi}}{q_m} g_m w(x) T_m(x), \tag{5}$$

where $\epsilon = ax + b$, and $\{\mu_{mi}\}$ are the Chebyshev moments of $\bar{\mathbf{H}}$ (discussed below),

$$\mu_{mi} = < i|T_m(\bar{\mathbf{H}})|i > . \tag{6}$$

The Gibbs factors $\{g_m\}$ in (5) are designed to reduce the Gibbs oscillations resulting from the finite-M truncation of the series. We choose the form derived by Jackson [11,17], which maintains the desired positive-definite nature of the DOS, while minimizing the maximum error between the exact DOS and the approximate DOS.

The heart of the KPM is the relationship between the approximate DOS in (5) and the exact DOS. This connection can be more clearly stated in ϕ space, where $x = cos(\phi)$, and the mth Chebyshev polynomial is simply $cos(m\phi)$. In this ϕ space, the approximate DOS is related to the exact DOS by a convolution with the *kernel polynomial*, a positive-definite approximation to a Dirac-delta function,

$$\delta_K(\phi) = \sum_{m=0}^{M} \frac{1}{2q_m} g_m cos(m\phi), \tag{7}$$

whose width is proportional to $1/M$. Thus, (5) represents a smearing of the exact DOS with a known resolution function. (In energy space, the width of the smearing function is not constant, reaching a maximum at $\epsilon = 0$.) The generality of the KPM allows, for example, smearing of the Fermi level instead of the DOS.

The Chebyshev moments in (6) are computed by defining the moment vector $\psi_m = T_m(\bar{\mathbf{H}})|i >$, such that $< i|T_m(\bar{\mathbf{H}})|i >= \mu_{mi} = \psi_0^\dagger \psi_m$. The main computational work comes

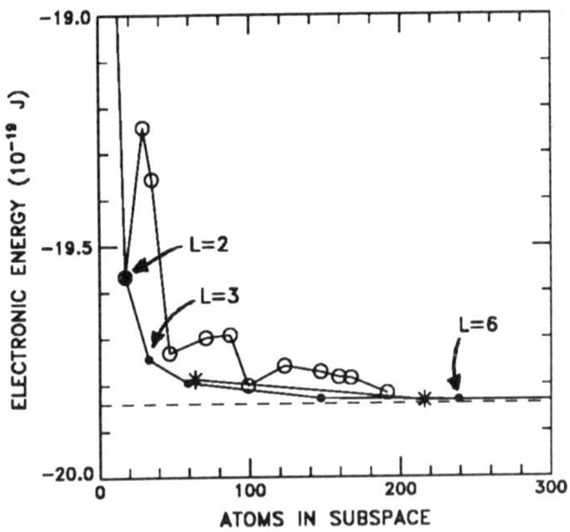

FIG. 1. Exact subspace E_{elec} (10^{-19} J) for bulk Si atom using physical truncation (∘), or logical truncation (•), compared to supercell calculations (*) and 1000-atom supercell (dashed line).

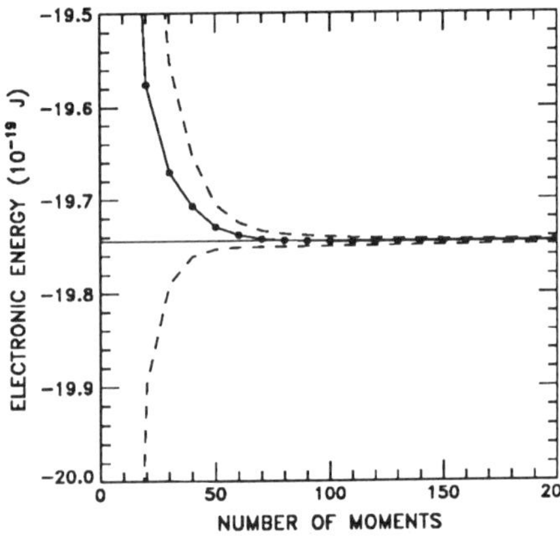

FIG. 2. Si E_{elec} using KPM with L=3 truncation; smeared Fermi level (upper dash), smeared DOS (lower dash), and average (solid). Straight line is exact L=3 result.

from the M matrix-vector multiplies using the recurrence relation, $\psi_m = 2\bar{\mathbf{H}}\psi_{m-1} - \psi_{m-2}$. Because $\mathbf{H}$ is sparse, each matrix-vector multiply requires computation proportional to the subspace size, N_i. This is then repeated for each basis function in the system, so the overall work scales as MN_iN, or MN^2 if there is no local truncation.

Differentiating the energy with respect to an atom coordinate requires the derivative of $< i|T_m(\bar{\mathbf{H}})|i >$. If no local truncation is applied, the derivative of the total moment $[\sum_i^N < i|T_m(\bar{\mathbf{H}})|i >]$ is straightforward; the trace over individual moments allows cyclic permutation of the matrices in each polynomial term, so that the scalar equivalent can be employed. In the case of local truncation, which breaks the trace, this approach is only approximate. To obtain the exact derivative requires a full matrix treatment of the polynomial expansion [18], which can be accomplished in a fashion that retains N scaling.

Figure 2 shows the moment convergence of the electronic energy (per atom) for the L=3 truncated Si system. The upper dashed curve results from applying the KPM to smear the Fermi function. This is an upper bound on the exact energy if E_F is exact. The lower dashed curve results from smearing the DOS. This is a lower bound if the smearing is uniform. The average of these two approximate bounds provides a good estimate of the exact energy, which we use hereafter. Convergence to the exact L=3 result is seen to be rapid, deviating by .08% at M=50 and .005% (3.8×10^{-22}J or 2.4×10^{-3} eV) at M=100.

Figure 3 shows the moment convergence of the unrelaxed vacancy formation energy to the exact L=3 result of 8.04×10^{-19}J (dashed line), using the atom-neutral method. Also shown is the vacancy formation energy after a full atomic relaxation (216-atom system) at each moment level. It appears converged by M=100, offering a good approximation to the untruncated, relaxed, exact TB result (6.24×10^{-19}J = 3.89 eV).

First-principles [19] and TB [14] studies have shown that the relaxed Si vacancy exhibits a Jahn-Teller distortion, in which the four neighbors of the vacancy site, initially second nearest neighbors to each other, pair up to form two shorter bonds. This behavior

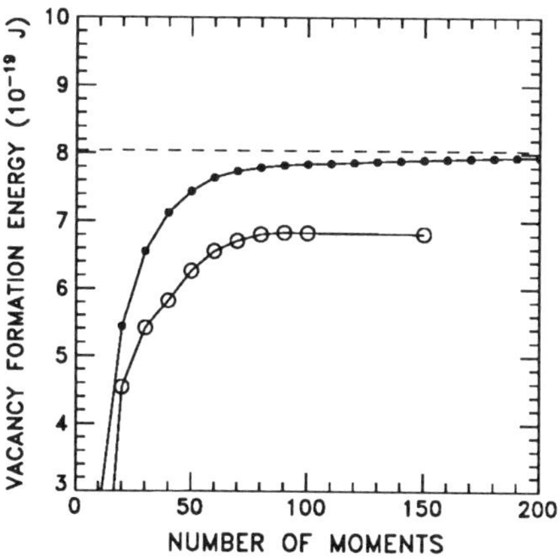

FIG. 3. Si vacancy formation energy using KPM with L=3 truncation; unrelaxed geometry (•) and relaxed geometry (o). The dashed line is the exact L=3 result (unrelaxed).

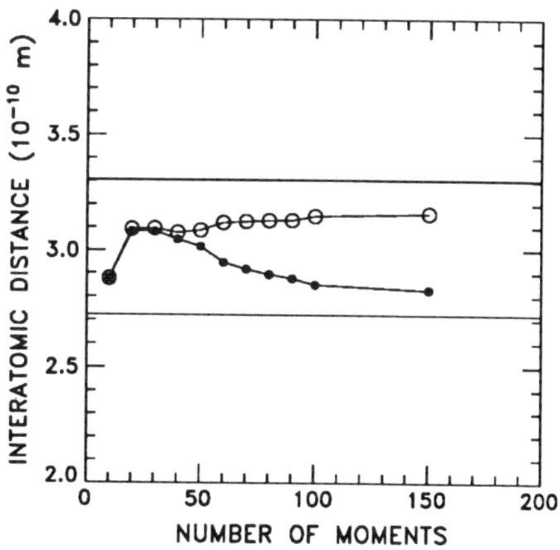

FIG. 4. Distances between atoms adjacent to relaxed vacancy, using L=3 KPM. Jahn-Teller distortion breaks R_1=R_2 symmetry, leading to short bonds [R_1(•)] and long bonds [R_2(o)]. Solid lines show R_1 and R_2 for untruncated, exact TB.

is properly predicted by the L-truncated KPM approximation if enough moments are used. Figure 4 shows the relaxed distance between the two unique pairs of vacancy-neighbor atoms as a function of M. Up to $M \sim 30$, the two distances are equivalent ($R_1 = R_2$), but reduced from the unrelaxed bulk value (3.84×10^{-10}m), indicating a symmetric relaxation of the atoms towards the vacancy. At $M \sim 50$, there is a bifurcation as the Jahn-Teller distortion occurs, with R_1, the distance between the atoms making the new bond, becoming shorter than R_2, the distance between the pairs of atoms not involved in a new bond. (Note that this new "bond" is still much longer than the bulk nearest-neighbor distance, 2.35×10^{-10}m). By $M = 100$, there is qualitative agreement with the exact, untruncated TB geometry (solid lines). The bifurcation at $M \sim 50$ is a consequence of the M-dependent energy resolution of the DOS. Below a critical value of M, the smearing obscures the information that there is a partially filled state at the Fermi level.

CONCLUSIONS AND ACKNOWLEDGMENTS

The local N-scaling approximation to TB presented here should be adaptable as an interatomic potential. A logical truncation range of L=3, with ~100 moments, appears adequate to obtain qualitative and quantitative features of defect energetics and geometries. This approach should also work for metallic and metallic-covalent systems, as the embedded atom method [20] corresponds to a second-moment (M=2) approximation to TB [21,22].

Work supported by the U.S. Department of Energy, Office of Basic Energy Sciences.

REFERENCES

[1] G. Galli, and M. Parrinello, Phys. Rev. Lett. **69**, 3547 (1992).
[2] X.-P. Li, R.W. Nunes, and D. Vanderbilt, Phys. Rev. B **47**, 10891 (1993).
[3] F. Mauri, G. Galli, and R. Car, Phys. Rev. B **47**, 9973 (1993).
[4] P. Ordejon, D.A. Drabold, M. Grunbach, and R. Martin, Phys. Rev. B **48**, 14646 (1993).
[5] M.S. Daw, Phys. Rev. B **47**, 10895 (1993).
[6] D. A. Drabold and O. F. Sankey, Phys. Rev. Lett. **70**, 3631 (1993).
[7] S. Goedecker and L. Colombo, Phys. Rev. Lett. **73**, 122 (1994).
[8] F. Mauri and G. Galli, Phys. Rev. B **50**, 4316 (1994).
[9] E.B. Stechel, A.R. Williams, and P.J. Feibelman, Phys. Rev. B **49**, 10088 (1994).
[10] R.N. Silver, H. Roder, A.F. Voter, and J.D. Kress, in *Simulation MultiConference '95 Proceedings, High Performance Computing '95*, edited by A. Tentner, p. 200 (1995).
[11] R.N. Silver, H. Roder, A.F. Voter, and J.D. Kress, J. Comp. Phys., submitted.
[12] A.P. Sutton, M.W. Finnis, D.G. Pettifor, and Y. Ohta, J. Phys. C **21**, 35 (1988).
[13] L. Goodwin, A.J. Skinner, and D.G. Pettifor, Europhys. Lett. **9**, 701 (1989).
[14] C.Z. Wang, C.T. Chan and K.M. Ho, Phys. Rev. Lett. **66**, 189 (1991).
[15] J.D. Kress and A.F. Voter, submitted to Phys. Rev. B.
[16] R.N. Silver and H. Roder, Int. J. Mod. Phys. C, **5**, 735 (1994).
[17] D. Jackson, *Theory of Approximations*, (American Mathematical Society Colloquium Publications, Vol XI, Providence, RI, 1930).
[18] A.F. Voter, R.N. Silver, and J.D. Kress, in preparation.
[19] G.A. Baraff and M. Schluter, Phys. Rev. B **30**, 3460 (1984).
[20] M.S. Daw and M.I. Baskes, Phys. Rev. Lett. **50**, 1285 (1983).
[21] M.W. Finnis and J.E. Sinclair, Phil. Mag. A **50**, 45 (1984).
[22] A.F. Voter, in *Intermetallic Compounds: Principles and Practice*, edited by J.H. Westbrook and R.L. Fleischer, John Wiley and Sons, Ltd, 1994), Vol. 1, p. 77.

AM1 AND PM3 SILICA STRUCTURE-ENVIRONMENT CALCULATIONS

Larry L. Hench and Jon K. West
Department of Materials Science and Engineering
University of Florida, Gainesville, FL 32611

Amorphous silica (a-silica) is modeled by 2-, 3-, 4-, 5-, 6-membered rings using AM1 and PM3 semi-empirical molecular orbital methods. Si-O bond lengths from 1.69 Å to 1.74 Å and Si-O-Si bond angles from 102° to 174° were obtained. 5- and 6-membered rings had minimal heats of formation. Infrared spectra from a mixture of 4-, 5, and 6-membered rings match well with experiment. Effects of strain and water on breaking Si-O bonds in the rings show quantitized paths for fracture involving ring contraction and hydrolysis. The MO models explain various environmental factors involved in the processing of sol-gel derived silica optical components.

INTRODUCTION

Understanding the relationship between the molecular structure and the physical and chemical properties of silica and silicates is one of the most significant topics in materials science because of the ubiquitous nature of silicon-oxygen containing compounds in the geosphere and biosphere.[1-3] However, efforts to correlate the molecular structure of silica and silicates with physical properties at a quantum mechanical level have been restricted due to limitations in the quantum mechanical models used to represent the electronic interactions in the SiO_2-containing clusters.[4-13] Many of the structures calculated involve only a few tetrahedra and siloxane bonds. Prior molecular orbital (MO) methods have also been limited in their ability to model simultaneous changes in bond angles and bond distances.[10] Calculations of transition states of activated intermediate complexes were also not possible for larger clusters until recently.

During the last three years the ease of modeling silicon-based compounds has improved dramatically. New semi-empirical MO methods have been produced by quantum chemists which are capable of calculating silica clusters with up to six tetrahedra per cluster in very reasonable time periods.[9] The MO models are sufficiently rapid that the transition states associated with environmental effects can be calculated. This has made it possible to achieve a quantitative theory of the fracture of silica and compute the interaction of biological molecules with silica surfaces. This paper reviews the calculational methods used in these studies and the results obtained for a range of SiO_2 structure-property-environment interactions.

CALCULATIONAL METHODS USED

Figure 1 is a schematic depicting some of the alternative molecular orbital (MO) calculational methods that are available. The cluster size and time for computation have a relationship to one another. The computational (CPU) time increases for *ab-initio* methods (STO-3G, 3-21 G, 6-31G*, 6-31G**, MP2, MP4) as well as for cluster size; i.e., number of atoms in a molecule or molecular cluster. The CPU time for *ab-initio* methods is too long to model silica-environment interactions. However, several semi-empirical methods, especially AM1 and PM3, give very good thermodynamic results with only a modest increase in calculation (CPU) time.[10] Both AM1 and PM3 permit calculation of molecular clusters large enough to model the effects of distribution of ring sizes, effects of strain and the inorganic-organic reactions of interest in modelling bioactivity factors. Errors increase dramatically for the Huckel methods even though very large collections of atoms may be modeled. Because of the errors we do not use Huckel methods except for a first approximation calculation of very large clusters. For very large structures, such as models of protein templates for epitaxial nucleation of biogenic silica, we use molecular mechanics models as a starting point.

In this review we discuss silica-based calculational models using a CAChe Worksystem®, Beaverton, OR. The models provide an interface to ZINDO [ZINDO] (M.C. Zerner's Intermediate Neglect of Differential Overlap program)[14,15] through a Macintosh environment as well as MOPAC [MOPAC] that provides an interface to MNDO AM1 and PM3 semi-empirical methods.

Davis and colleagues have shown that the modified neglect of diatomic overlap (MNDO) semi-empirical MO method can be used to model the reactions of Si-O containing molecules.[11,13] However, there are two situations in which MNDO produces large errors. MNDO overestimates core-core repulsions between atoms when they are separated by approximately van der Waals distances. This means that MNDO cannot reproduce H-bonds or accurately calculate the heat of formation, Hf, for strained or crowded

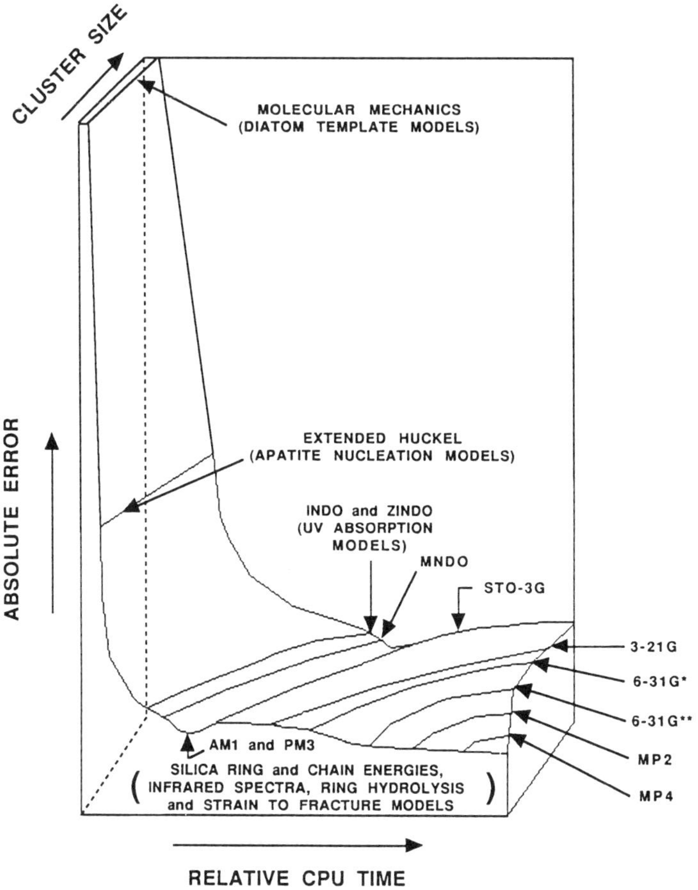

Figure 1. Effect of molecular cluster size on computational time (CPU) and calculational error.

molecules, such as those involved in bond fracture, or many transition state structures. The MNDO method also overestimates transition state energies and distorts transition state geometries, again fatal flaws for calculating the interaction of organic-inorganic clusters. The Austin Method (AM1) [MOPAC][16-18] semi-empirical method reduces these problems by modifying the core repulsion function used in MNDO with additional Gaussian terms and the calculated structures and heats of formation match quite well with experimental values.[10] ZINDO provides two different valence-electron-only semi-empirical procedures: a method for computing spectroscopic properties (electronic spectra) and a method of computing molecular geometries (conformations and structures).[14]

ZINDO uses a theoretically-based INDO parameterization and therefore contains parameters for more elements than does the experimentally-based MOPAC.[14] Thus, ZINDO can be used to calculate geometries for molecules which cannot be calculated with MOPAC. However, because MOPAC has been specifically parameterized for groundstate geometries and heats of formation, MOPAC (with the AM1 or PM3 parameters) is the better MO method for computing molecular structures and energetics. ZINDO can use d orbitals, and therefore can be used for calculations involving most transition series metals. By contrast, MOPAC uses only s and p orbitals and cannot be used for transition metals. ZINDO has an alternative parameterization for electronic (UV/visible) spectra named INDO/S which produces good results. Since MOPAC is parameterized for ground-state heats of formation, excitation energies determined by MOPAC will probably not be as good as those from ZINDO. Although the ZINDO geometry optimization is faster than MOPAC, the geometries obtained are not as reliable as those from most MOPAC models (MNDO, AM1 and PM3). ZINDO geometries are nearly as accurate as the MOPAC MINDO/3 geometries.[14]

As a calculational strategy, the geometries for silicate rings and chains and environmental effects discussed herein were optimized using AM1 or PM3. These optimized geometries were then submitted to ZINDO configuration interaction (CI) calculations to predict the electronic spectra. The same AM1 geometries were also submitted to a FORCE calculation within MOPAC to calculate the infrared spectra.

STRUCTURAL MODELS

Various mechanical, x-ray, ab-initio and semi-empirical quantum molecular orbital (MO) methods have been used to investigate the geometries of silica rings of different sizes and the effect of ring size on bond energy and infrared and Raman vibrational modes.[3-8,19-23]

We[19-21] have used both AM1 and PM3[24,25] methods to calculate optimized structures for 2-, 3-, 4-, 5-, and 6-member silica rings. The two models produce similar lowest energy geometries for small rings (2, 3 and 4 tetrahedra) but larger rings modeled by PM3 have different symmetry than AM1 rings. AM1 predicts that the 5-member ring is the most stable while PM3 predicts that the 6-member ring is more stable (Fig. 2). The Galeener ab-initio study[7] predicts the 4-member ring to be most stable. The ring size distribution from the Bell and Dean[22] model of a-SiO_2 is also shown in Fig. 2. For details of these results see Refs. 20 and 21.

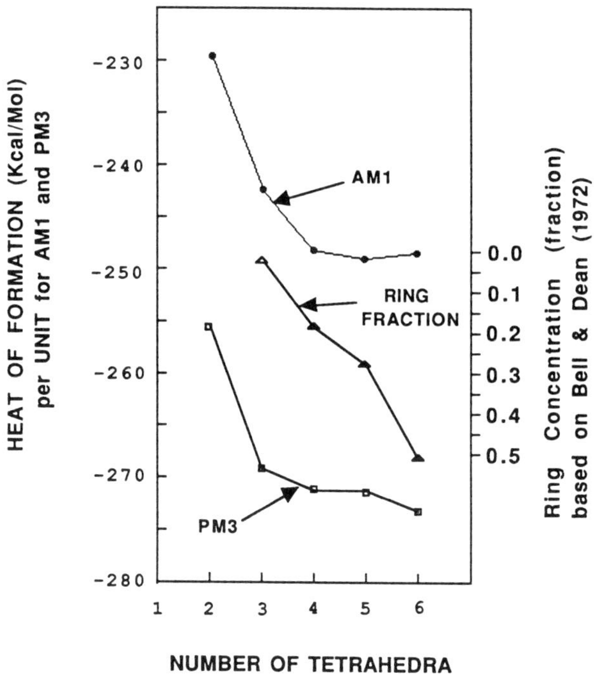

Figure 2. Distribution of energies for AM1, PM3 and Bell and Dean silica ring models.

OPTICAL SPECTRA

Both AM1 and PM3 methods can be used to calculate force constants between all neighboring atoms which are converted to a vibrational spectrum using a FORCE, normal mode analysis, calculation with MOPAC.[24,25] The frequency of vibrational transitions are predicted by the calculations that compute the force necessary to distort atom pairs in the molecular cluster from their equilibrium geometries. The intensities of the infrared active transitions are plotted. The vibrational spectra obtained in this manner are idealized, gas phase spectra and therefore are only an approximation to a-silica properties. Also molecular transitions which are forbidden by symmetry in the MO models are present in experimental spectra of the condensed phase spectra. Anharmonicity effects, overtones, or combination transitions are also absent from the theoretical IR spectra. Thus, the calculated IR spectra are sharp, discrete lines whereas experimental a-SiO_2 modes are broadened by rotational levels, instrument optics, molecular motions, and other solid state effects. These effects are approximated in the calculated spectra by introducing a gaussian-like distribution for each vibrational mode with a half width of 50 cm^{-1}.

Figure 3 shows the experimental IR spectra for a-silica with assignments of the major vibrations.[26] Figure 3 also shows the results of both AM1 and PM3 models for the 4-member silica rings. Calculated IR spectra from both models for 5-member and 6-member rings are very similar to the spectra in Fig. 3. The theoretical IR spectra obtained from the AM1 and PM3 calculations provide a much closer match to the experimental a-SiO_2 spectrum than earlier calculations.[8] The difference between theoretical and

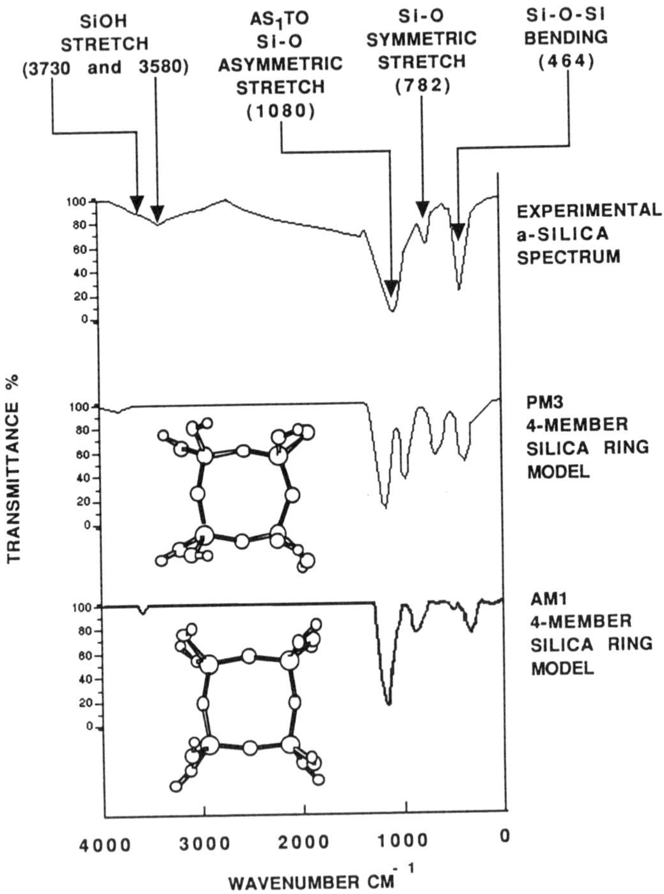

Figure 3. IR spectra for silica, PM3 4-member ring and AM1 4-member ring.

experimental values for the 1080 cm^{-1} Si-O ASTO stretch is small for both MO models. The AM1 spectra are shifted approximately 30 cm^{-1} to higher frequencies compared to experiment. The PM3 results show an extra vibration near 500 cm^{-1}.

Chains of silica tetrahedra containing 3-, 4-, 5- and 6-tetrahedra were also modeled using AM1 and ZINDO.[27] The chain structures and equivalent ring structures[21] were submitted to ZINDO configuration interaction (C.I. = 6) calculations. The results in Fig. 4 show that there is little or no change in the UV absorption edge of the silica chain structures regardless of chain lengths.

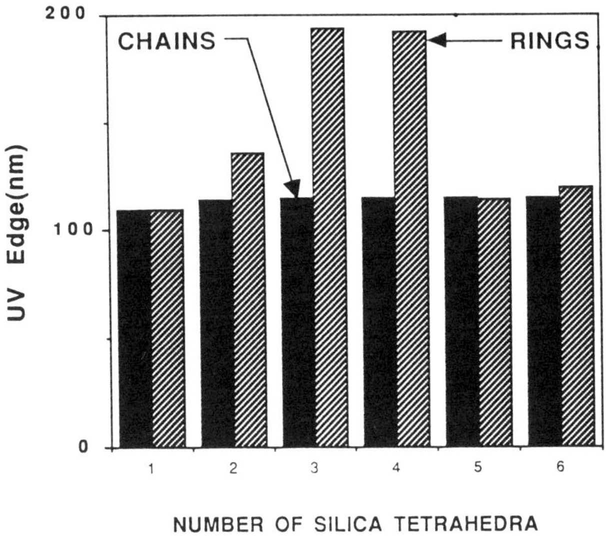

Figure 4. Comparison of the theoretical electronic UV edge between silica rings and chains.

In contrast, the MO models of silica rings show (Fig. 4) a dramatic change of nearly 80 nm in the UV absorption edge. Figure 5 shows the calculated electronic spectra for a 4-member AM1-ZINDO ring, a PM3-ZINDO ring, porous gel-silica[27] and fully dense gel-silica.[27] The calculations predict that a maximum (longer wavelength) absorption should occur for materials with a large concentration of 3-fold and 4-fold rings[27] Wallace, using RAMAN spectroscopy, has shown that porous Type VI gel-silica shows a change in the concentration of 3-member and 4-member rings with processing temperature[28] and with water content.[29] The gel-silica network density also

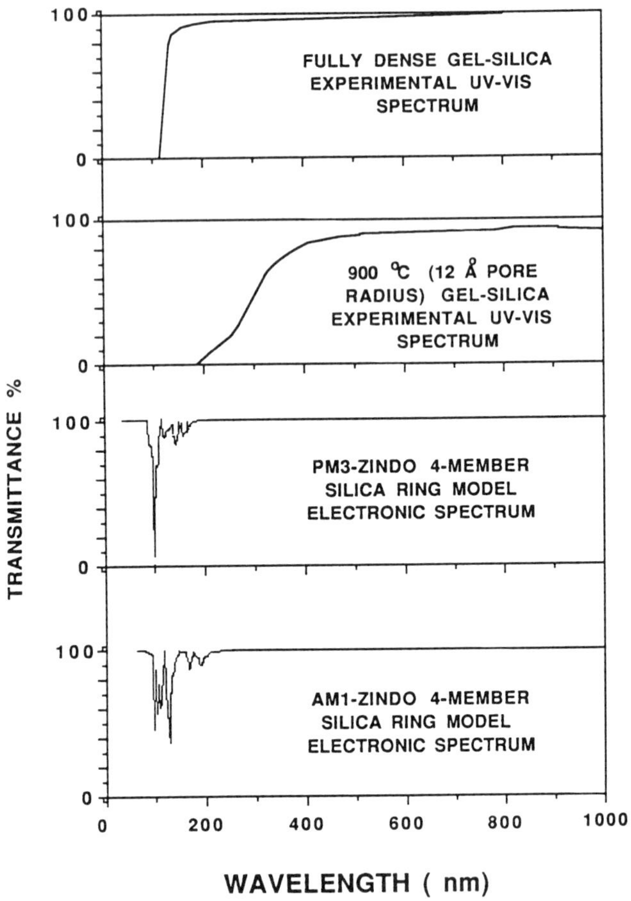

Figure 5. Electronic spectra for fully dense silica, porous silica, PM3 ZINDO 4-member, AM ZINDO 4-member ring.

increases as processing temperature is increased.[3,28,30] Figure 6 shows that the 50% UV cutoff also goes through a maximum (longer wavelength) as the silica gel-glass processing temperature increases because of structural changes. This effect is probably not seen in small pore (1.2 nm) radius gels because the [OH] is much higher than for larger pore gels.[3,27,31]

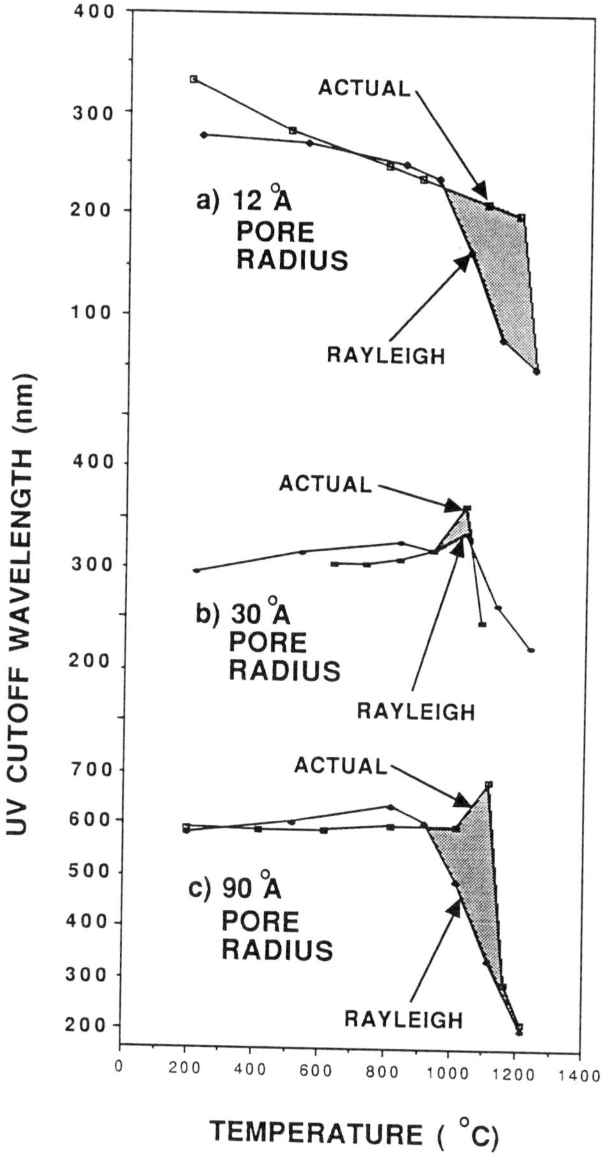

Figure 6. UV cutoff as a function of processing temperature of porous gel-silica.

We propose the following equation as a first approximation to predict the wavelength, λ, cutoff of the UV cutoff in porous Type VI optical silica:

$$\lambda_{cutoff} = \alpha[OH] + \beta[D_2] + \gamma[D_1] + \delta(\Delta)^{+3/2} \qquad (1)$$

where [OH] is the concentration of silanols $[D_2]$ = concentration of 3-member silica rings, $[D_1]$ = concentration of 4-member silica rings, Δ = network wall thickness (Rayleigh scattering component), and α, ß, γ, δ are linear coefficients. The difference between the calculated Rayleigh scattering and the experimentally measured scattering has been shaded in Fig. 6. The excess UV absorption is likely to be due to the large number of 3 and 4 member rings in porous gel-silica.

The effect of 3-member and 4-member rings on UV absorption is not seen in bulk silica because of the very low concentrations of these structural units as compared to the porous materials.

SILICA FRACTURE

It has been recognized[32] since the time of Inglis[33] and Griffith[34] that fracture occurs on an atomistic level when bonds between atoms are strained past the breaking point. When a tensile stress at the tip of a crack exceeds a critical value crack growth occurs.[32] This concept underlying fracture is typically modeled as a continuous stress-displacement curve, approximated as a sine curve. However, because of structural constraints and the finite size of molecular bonds, a crack tip must advance in increments with the size of the step dictated by the localized molecular structure containing the crack tip. A continuum of strain applied to a material must be translated mechanistically into quantitized steps of bond breaking.[35]

Molecular models have been proposed for the environmental effects on fracture[36,37] but they do not describe quantitative reaction pathways for the fracture process. We have used AM1 to calculate the reaction pathways for siloxane bond fracture under simulated vacuum[35] and hydrolysis[38] conditions. Step wise strains of 0.02 nm were applied to 3-member, 4-member and a 5-member ring-chain structure.[35] The fracture barrier for 5-member ring-chain structures is 103 Kcal/mol. The barrier for fracture of a 3-member ring is +96 Kcal/mol. Fracture by contraction of 4-member rings has a lower energy barrier of +77 Kcal/mol due to formation of pentacoordinate silicon transition states which produce trisiloxane rings and a broken siloxane bond (Fig. 7). thus, the ring contraction model predicts

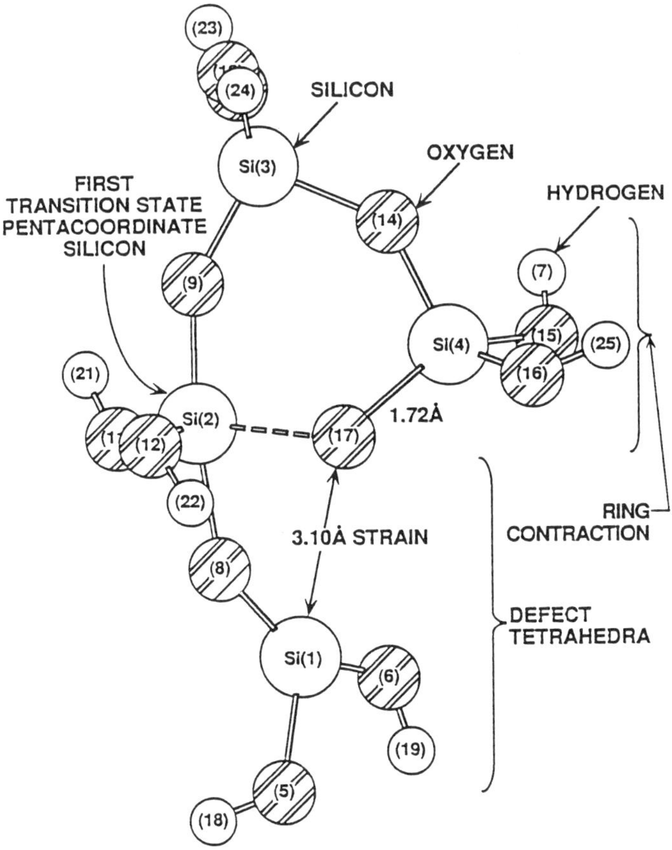

AM1 RHF HEAT OF FORMATION = -923.4 KCAL/MOL
(MODEL BB11)

AM1 UHF HEAT OF FORMATION = -925.9 KCAL/MOL
(MODEL BB36)

Figure 7. First transition state for Si-O bond fracture without water in a 4-membered silica ring.

that a crack will follow a path which depends on the distribution of 4-member (or larger) rings in vacuum or fast fracture.[38]

Using the same AM1 method, the transition state has been found for the contraction of a 5-member ring into a 4-member ring. The activation barrier to fracture by this contraction is only $E_f = +14.82$ Kcal/mol.

Thus, silica fracture in vacuum via ring contraction yields lower fracture energies than expected from simple Si-O bond breaking. Several different schemes of ring contractions are possible for both 5-member and 6-member ring structures. All contraction paths have different intermediate structures that lead to the same end point of crack growth. The various barriers to fracture range from +8 to +52 Kcal/mol for the 5-member ring contractions and from +9 to +41 Kcal/mol for 6-member ring contractions. This step-wise, quantitized contraction mechanism for crack propagation is illustrated in Fig. 8.

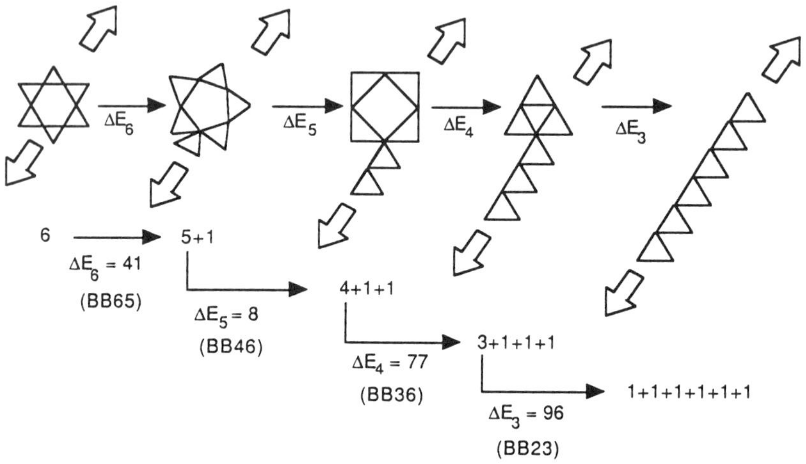

AM1 UHF ENERGY (KCAL/MOL)

Figure 8. Simplified ring contraction schemes during a-silica fracture in vacuum.

Hydrolysis of the ring structures can also occur if a polar molecule like water attacks a strained silicon-oxygen bond.[38] A quantitative model of environmentally sensitive crack growth of amorphous silica (a-silica) is based upon calculations (AM1 method) of a water molecule interacting with strained 3-member and 4-member silica rings and a 5-member ring-chain structure. The results[38] show very much lower energy barriers than fracture without water present. The energy barrier for hydrolysis of strained 3-member rings is only +7 Kcal/mol, the energy barrier for hydrolysis of strained 4-member rings is +29 Kcal/mol, and for a 5-member ring-chain structure it is +39 Kcal/mol. The reaction path for fracture of the Si-O bond of a 4-member ring by interaction with water is shown in Fig. 9.

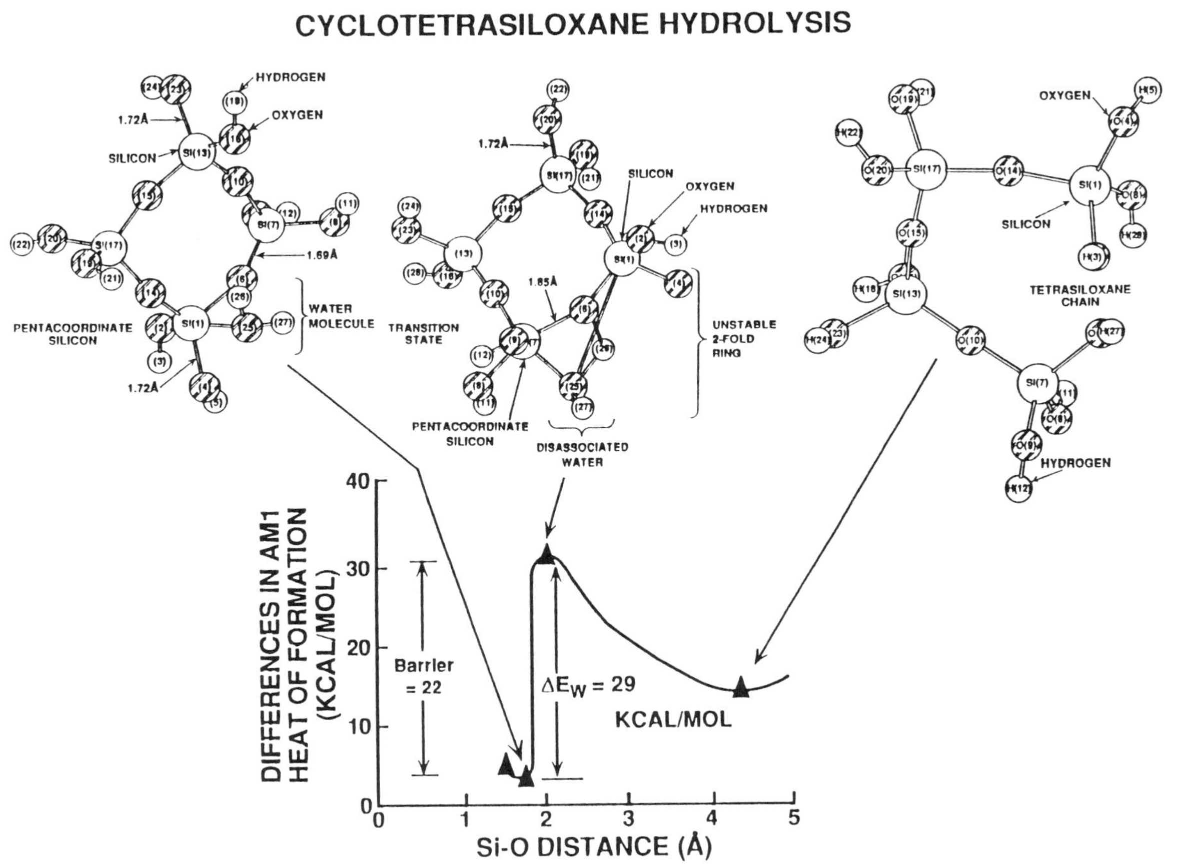

Figure 9. Reaction path for Si-O bond fracture with interaction of water for a 4-membered silica ring.

Thus, the MO model predicts that the energetics of Region 1 slow crack growth[40] is controlled primarily by the distribution and hydrolysis of 3-membered silica rings in the a-silica structure. Region III crack growth is controlled by the distribution and energy of contraction of 4, 5 and 5+ membered rings. The quantum mechanical origin of the difference in the regions of v-K behavior is summarized in Fig. 10; i.e. there are substantially lower energy barriers when H_2O molecules are present.

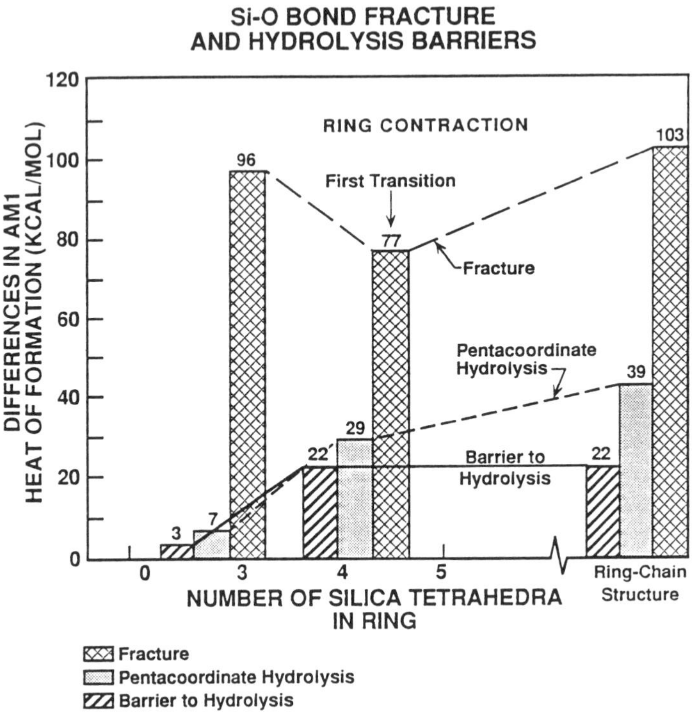

Figure 10. Differences in energy barriers (AM1 method) to fracture and hydrolysis for different size silica structures.

This step wise contraction mechanism for crack propagation can be used to estimate the crack velocity as a function of stress intensity, K_1, using the equation:[40]

$$V = V_o \exp\left[\frac{-E_a + \beta K_1}{RT}\right] \qquad (2)$$

Table I shows a weighted average of the fracture barriers that can be estimated from the Bell and Dean[22] distribution of sizes of silica rings in a-SiO_2. The concentration of 3-member rings is estimated from Wallace's Raman studies.[29] A weighted average of the fracture barrier, E_a, is calculated as the sum of the product of ring fracture times the corresponding fracture barriers for the rings.

Table I. Weighted Average Fracture Barrier Based Upon Silica Ring Distribution.

Bell and Dean	3-member	4-member	5-member	6-member
Number	7	48	73	137
Fraction	0.026	0.181	0.275	0.517
E_a(Fracture Barrier) "Kcal/Mol"				
Vacuum	96	77	8	41
Water	3.9	22	22.6	22.6
Weighted Average	E_a (Vacuum) = 39.6			
	E_a (Water) = 22.0			

The weighted average barrier for fracture in vacuum is 39.6 Kcal/Mol and it is 22.0 for silica crack growth in water. Using V_o = 400 m/s and ß = 25 for both vacuum and water we get the v-K results as shown in Fig. 11.

CONCLUSIONS

This analysis of the quantum mechanical basis of the electronic spectra and fracture mechanics of a-SiO_2 yields results surprisingly close to experimental values. It is reasonable that molecular structure at the atomic level controls the details of electronic and molecular vibrations. It is also intriguing that the same MO models yield fracture mechanics relations that are also very close to experimental observations. Further studies will explore the effects of different environmental variables in an effort to compare MO predictions with experiment.

ACKNOWLEDGEMENTS

The authors gratefully acknowledge financial support of Air Force Office of Scientific Research Grant No. F49620-92-J-0351. We also acknowledge the technical advice of CaChe Scientific, Inc., Beaverton, OR, especially Sam Cole and George Pervis.

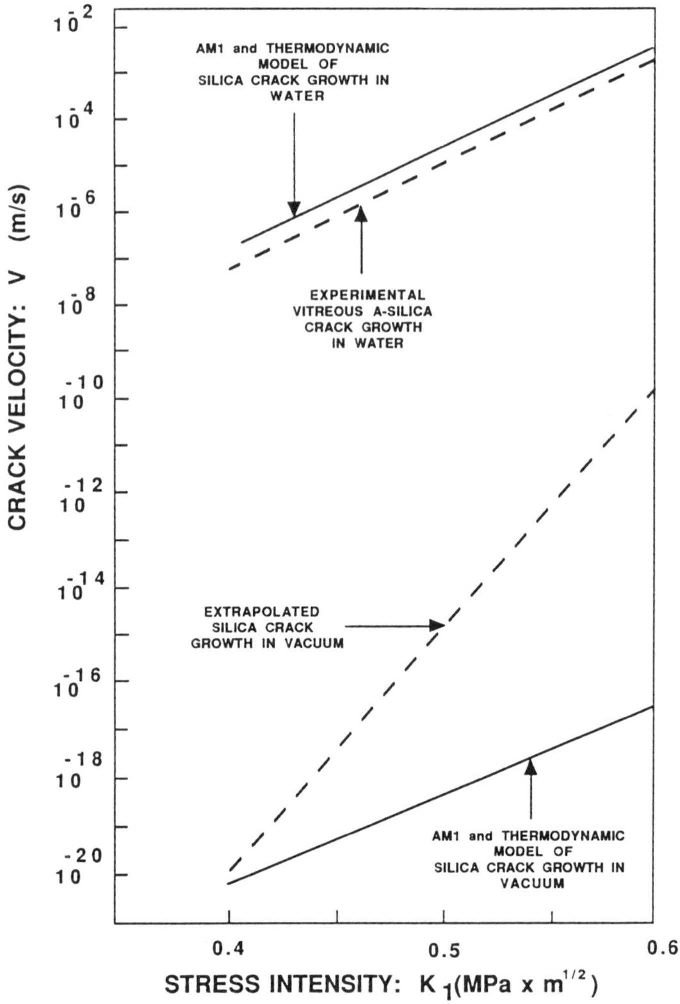

Figure 11. Comparison of theoretical and experimental U-K curves.

REFERENCES

1. R. K. Iler, *The Chemistry of Silica*, Wiley, New York, 1979.
2. D. Evered, ed., *Silicon Biochemistry*, Wiley, New York, 1986.

3. C. J. Brinker, G. W. Scherer, *Sol-Gel Science*, Academic Press, San Diego, 1990.
4. A. G. Revesz and G. V. Gibbs, In *The Physics of MOS Insulators*, ed. G. Lucovsky, S. T. Pantelides, F. L. Galeener, eds., Pergamon, New York, 1980, pp. 92-96.
5. G. V. Gibbs, Am. Mineralogist 67, 421-50, 1982.
6. A. C. Hess, P. F. McMillan and M. O'Keeffe, J. Chem. Phys. 90, 5661, 1986.
7. F. L. Galeener, J. Non-Crystalline Solids 49, 53, 1982.
8. N. Uchida and M. Shinmei, J. Non-Crystal. Solids 122, 276-84, 1990.
9. L. P. Davis and L. W. Burggraf In *Science of Ceramic Chemical Processing*, L. L. Hench, D. R. Ulrich, eds., Wiley, New York, 1986, pp. 400-411.
10. D. B. Boyd In *Reviews in Computational Chemistry,* K. B. Lipkowitz, D. B. Boyd, eds., VCH, New York, 1990, pp. 321-354.
11. L. P. Davis and L. W. Burggraf In *Ultrastructure Processing of Advanced Ceramics*, J. D. Mackenzie and D. R. Ulrich, eds., Wiley, New York, 1988, p. 367.
12. W. Wong-Ng, G. S. White and S. W. Freiman, J. Am. Ceram. Soc. 75(11), 1992, 3097-3102.
13. L. P. Davis, L. W. Burggraf and M. S. Gordon J. Amer. Ceram. Soc. 110, 1988, 3056-3062.
14. M. C. Zerner In *Review of Computational Chemistry II*, K. B. Lipkowitz and D. B. Boyd, eds., VCH, New York, 1991. pp. 313-66.
15. *ZINDO Version 3.0*, CaChe Scientific, Inc., Beaverton, OR
16. *MOPAC Version 6.1*, Tektronix, Inc., CaChe Scientific, Beaverton, OR
17. M.J.S. Dewar, E. G. Zoebisch, E. F. Healy and J. P. Stewart, J. Am. Chem. Soc. 107, 1985, 3902-3909.
18. M. S. Dewar and C. Jie, Organometallics 6, 1987,1486-1490.
19. T. Chia, J. K. West and L. L. Hench In *Better Ceramics Through Chemistry VI*, A. K. Cheetham, C. J. Brinker, M. L. Mecartney, and C. Sanchez, Materials Research Society, Pittsburgh, Vol. 346, 1994, pp. 727-732.
20. J. K. West and L. L. Hench, J. Non-Crystal. Solids 180, 1994, 111-116.
21. J. K. West and L. L. Hench, J. Amer. Ceramic Soc., 1995, In press
22. R. J. Bell and P. Dean, Phil. Mag. 25, 1962, 1381-1398.
23. R. L. Mozzi and B. E. Warren, J. Appl. Crystallogr. 2, 1969, 164.
24. J.J.P. Stewart, J. Computational Chem. 10(2), 1989, 209-221.
25. J.J.P. Stewart, J. Computational Chem. 11(4), 1990, 543.

26. C. T. Kirk, Phys. Rev., B38, 1988, 1255.
27. J. K. West, G. P. LaTorre and L. L. Hench, J. Non-Crystal. Solids. In press 1995.
28. S. Wallace, *Porous Silica Gel Monoliths: Structural Evolution and Interaction with Water*. PhD thesis. Univ. Fla., Gainesville, 1991.
29. S. Wallace, J. K. West, and L. L. Hench, J. Non-Crystal. Solids 152, 1993, 101-108.
30. W. L. Vasconcelos, *Topological Evolution and Properties of Sol-Gel Silica Monoliths*, Ph.D. thesis. Univ. Fla., Gainesville, 1989.
31. L. L. Hench and J. K. West, Chem. Rev. 90, 1990, 33-72.
32. S. W. Freiman In *Glass: Science and Technology*, D. R. Uhlmann and N. J. Kreidl, eds, Academic Press, New York, 1980, pp. 21-78.
33. C. E. Inglis, Trans. Inst. Nav. Arch. 55, 1913, 219-229.
34. A. A. Griffith, Trans. R. Soc. London 221, 1921, 163-198.
35. J. K. West and L. L. Hench, J. Maters. Sci. 29, 1994, 3601-3606.
36. T. A. Michalske and S. W. Freiman, J. Am. Ceram. Soc. 66, 1983, 284.
37. T. A. Michalske and B. C. Bunker, J. Appl. Phys. 56(10), 1984, 2686-2693.
38. J. K. West and L. L. Hench, "Silica Fracture, Part II: A Ring Opening Model Via Hydrolysis," J. Materials Science, 29, 1994, 5808-1816.
39. J. K. West and L. L. Hench, "Silica Fracture, Part III: 5-Fold and 6-Fold Silica Ring Contraction Models," J. Materials Science, 1995.
40. S. M. Wiederhorn, J. Amer Ceram. Soc. 50, 1967, 407-414.

MOLECULAR DYNAMICS SIMULATION AND NEUTRON DIFFRACTION STUDY OF ALKALI SILICATE GLASSES

B. Vessal, Catalysis and Sorption Project, Biosym Technologies, Inc., San Diego, CA

A.C. Wright, J.J. Thomson Physical Laboratory, Reading University, Whiteknights, Reading, UK

A.C. Hannon, ISIS Facility, Rutherford Appleton Laboratory, Chilton, Didcot, UK

ABSTRACT

Results have been presented for the molecular dynamics simulation of three alkali silicate glasses undertaken at the same exact composition as those studied by neutron diffraction. These simulations have been tested by rigorous comparison with the experimental data. The results demonstrate the powerful role that molecular dynamics can play in interpreting diffraction data for multicomponent glasses.

INTRODUCTION

The traditional glass industry is for the most part based on silicate glasses but, for materials which have been utilised for several millennia, surprisingly little is known about their detailed structure, mainly due to the non-unique nature of diffraction studies of amorphous solids and the consequent need to rely on modeling techniques such as molecular dynamics to extract three-dimensional structural information from one-dimensional diffraction data. Of particular interest is the so-called mixed alkali effect [1], which involves the extremely non-linear behaviour of the properties of glasses containing two or more modifier cations and at present is particularly poorly understood.

The present study aims to provide accurate neutron diffraction data with high real space resolution for a series of alkali silicates and to interpret these data using state of art molecular dynamics simulations. The glasses investigated includes several compositions from the Na_2O-SiO_2 system which is the archetypal network modifier- network former combination, and two further samples; vitreous K_2O-SiO_2 and the glass containing approximately equal amounts of Na_2O and K_2O to study the structural origins of the mixed alkali effect. The results from initial neutron diffraction measurements have already been published, together with a more detailed literature review [2], and these have now been extended to much higher values of the scattering vector, Q, to achieve the necessary increase in real space resolution to investigate the detailed geometry of the basic SiO_4 structural units. This is extremely important for checking the validity of molecular dynamics simulations, since earlier simulations using 2-body potentials have always yielded structures in which the tetrahedral distortion is far too great. The present paper reports the new higher Q data for three of the glasses together with preliminary results from molecular dynamics simulations using

3-body potentials, which are being performed at compositions as close as possible to those studied by diffraction and tested by rigorous comparison with the experimental data [3].

TECHNIQUES

Full details of the sample preparation are given, together with an account of the earlier measurements, in reference [2] and the compositions of the glasses discussed here are summarized in Table I. The diffraction data were extended to higher Q [38A^{-1}] using the LAD spectrometer at the ISIS spallation neutron source and the resulting correlation functions are illustrated in Figure 1, normalised to a composition $xM_2O\text{-}SiO_2$ [M=Na, K]. Constant volume molecular dynamics simulations were undertaken [c.f. Table I] with the simulation cell of volume 7538.9A^3. The starting structures were generated from that of crystalline $\alpha\text{–}Na_2O.2SiO_2$ [4], the appropriate number of atoms being removed, using BIOSYM software, to obtain both charge neutrality and the closest agreement with the experimental composition. The crystalline structure was melted at 6000 K followed by quenching to 2500 K, 1250 K, 625 K, and 293 K. The system was run for 2500 steps at each temperature. The timestep used for integrating the equations of motion was 3 fs. A molecular graphics representation of the $Na_2O\text{-}SiO_2$ glass structure is shown in Figure 2.

Table I. Experimental and Simulation Details [ρ^0 is given in units of 10^{-2} atoms A^{-2}. The first four columns under the simulation heading list the number of different ions ued in generating the models.].

	Experiment			Simulation					$R\chi$ [%]
Sample	x_{Na_2O}	x_{K_2O}	ρ^0	Na	K	Si	O	ρ^0	
N5	0.446	-	7.343	114	-	128	313	7.362	12.7
NK	0.229	0.235	6.837	54	54	117	288	6.805	12.9
K	-	0.514	6.165	-	106	102	257	6.168	14.1

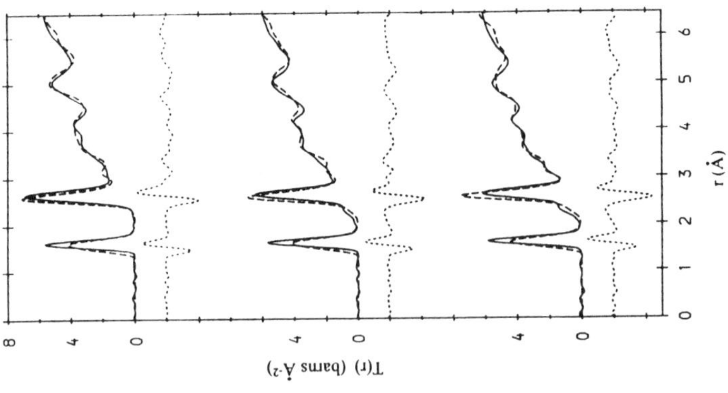

Figure 1 The real space correlation function for $K_2O\text{-}SiO_2$ (top), $Na_2O\text{-}K_2O\text{-}SiO_2$ (middle), and $Na_2O\text{-}SiO_2$ (bottom). Solid lines, neutron diffraction ; dashed lines, molecular dynamics and dotted lines, difference curves (experiment - simulation).

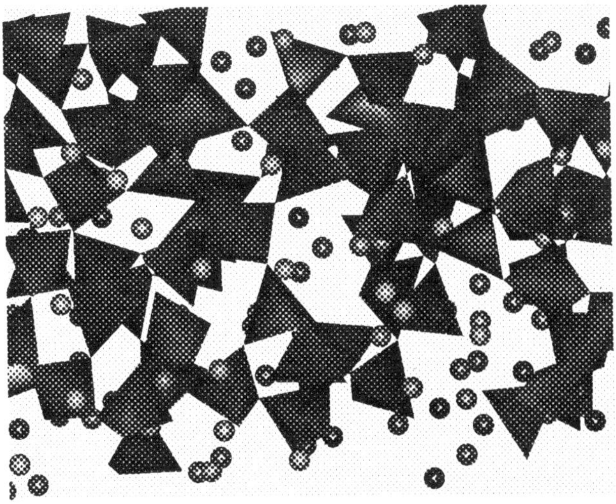

Figure 2 The structure of Na_2O-SiO_2, generated using BIOSYM software, showing the SiO_4 tetrahedra and Na^+ ions as spheres.

Simulated glass structures are prepared computationally in the same way as real glasses. Models of crystalline systems are melted; the melts are then quenched freezing the structure into a disordered glassy phase. The main problem with this technique is in the time scale of the melting and more particularly of the quench, as with current computational capabilities, these are several orders of magnitude greater than the rates achieved experimentally. As a consequence, simulated glass structures have very high fictive temperatures. Nevertheless, the information obtained on glass structures is of great value.

Now we give the details of the interatomic potentials that were employed in our simulations.

The long range coulomb part of the potential is calculated using Ewald's method [5] , and employing full ionic charges on the silicons , oxygens and the alkali atoms. Following Sangster and Dixon [6], and Anastasiou and Fincham [7] we write the coulomb part of the potential as follows:

$$E^c = E^c_1 + E^c_2 + E^c_3 \qquad (1)$$

$$E^c_1 = \sum_{i=1}^{N} \sum_{\substack{j=1 \\ i>j}}^{N} Z_i Z_j \,\mathrm{erfc}(\alpha r_{ij}) / r_{ij} \qquad (2)$$

$$E_2^c = \sum_{i=1}^{N} Z_i^2 \alpha \pi^{-1/2} \qquad (3)$$

$$E_3^c = \sum_{i=1}^{N} \sum_{\mathbf{k}=\mathbf{0}} A(k) \left[\left(\sum_{j=1}^{N} Z_j e^{i\mathbf{k}.\mathbf{r}_j} \right) Z_i e^{-i\mathbf{k}.\mathbf{r}_i} \right] \qquad (4)$$

where E_1^c and E_2^c are known as the direct and the reciprocal lattice parts of the coulomb energy respectively. N is the number of ions in the simulation box, Z_i and Z_j are the charges on the ions i and j, r_{ij} is the distance between i and the nearest image of j, and α is called the Ewald parameter. The choice of this parameter determines the value of the direct and reciprocal sum cutoffs. If one wishes to determine the direct and reciprocal lattice sums to an accuracy A_c, then the value of the direct lattice cutoff that is used is $r_{max} = f/\alpha$, and the reciprocal lattice cutoff is $k_{max}/2\pi = \alpha f/\pi$, where $f = (-\ln A_c)^{1/2}$. We used a value of α that gives comparable number of terms in both the direct and reciprocal lattice sums, i.e.,

$$\alpha = (N\pi^3/V^2)^{1/6} \qquad (5)$$

where V is the volume of the simulation box.

The sum over **k** need only be carried out over half of reciprocal space and the final result doubled since reciprocal space is always centrosymmetric. Thus the z component of the reciprocal lattice vector will always be taken as positive.

In equation (4) **k** is 2π times a reciprocal lattice vector and **r** is an atom position. The maximum number of reciprocal lattice vectors in the x, y and z directions N_x, N_y, N_z are determined from the formula given below:

$$N_x = (\mathbf{k}_{max}/2\pi)\,|\mathbf{l}_x| \qquad (6)$$

where $\mathbf{l}_x$ is the x column of a matrix whose columns are the unit cell lattice vectors. A(**k**) is defined as

$$A(\mathbf{k}) = 2\pi/V \exp(-k^2/4\alpha^2)/k^2 \qquad (7)$$

We have not included the **k** = **0** term in our calculations.
A four range Buckingham potential is used to model the short range interactions between different ions.

$$E_{ij} = A_{ij} \exp(-r_{ij}/\rho_{ij}), \quad r_{ij} < r_1 \qquad (8)$$

$$E_{ij} = \sum_{m=0}^{5} A_m r_{ij}^m, \quad r_1 < r_{ij} < r_2 \qquad (9)$$

$$E_{ij} = \sum_{m=0}^{3} B_m r_{ij}^m, \quad r_2 < r_{ij} < r_3 \qquad (10)$$

$$E_{ij} = -C_{ij}/r_{ij}^6, \quad r_3 < r_{ij} < r_c \qquad (11)$$

where r_c is the short range cutoff. We have used a value of 7.6 A for r_c.

The function is splined at r_1, r_2, and r_3 so that the energy, and first and second derivatives are continuous. The function has a minimum at r_2. Despite the complex nature of the potentials, we are able to fit the different components using a cubic spline which is then used in the forces calculation.

A cubic spline is either fitted separately to the Coulomb direct space error function term and to the short range potential, or to the sum of the two terms. Two separate splines are needed if pressure is to be calculated. One thousand points are used in the spline. The spline is fitted to the square of the distance to avoid calculating a square root.

An essential feature of our potential model is the inclusion of two different types of three-body potentials, one between the O-Si-O triads and the other between the Si-O-Si triads. We have experimented with different forms of three-body interactions in the simulation of vitreous silica at constant volume. We find the following form to be quite effective in reproducing the right BAD for both O-Si-O and Si-O-Si angles.

$$E_{ijk} = k_{ijk} [\, \theta^n (\theta-\theta_0)^2 (\theta+\theta_0-2\pi)^2 - n/2\pi^{n-1} (\theta-\theta_0)^2 (\pi-\theta_0)^3 \,] \qquad (12)$$

and where k_{ijk} is the three-body spring constant and θ_0 is the equilibrium bond angle, i.e. $109^o 28'$ in the case of O-Si-O interactions and 144° for Si-O-Si interactions, and θ is the calculated bond angle.

The parameters for the short-range and the three-body potentials are determined using the THBFIT code which uses a least-squares fitting method to get the best agreement between the observed and the computed properties of α-quartz. These parameters are listed in Table II.

Table II. Potential Parameters

Parameter	Si-O	O-O	Na-O	K-O	Units
A_{ij}	990.617	4511887.2	611.10	902.79	eV
ρ_{ij}	0.3297	0.149	0.3535	0.3698	A
C_{ij}	25.0	52.12	0.0	0.0	eVA^6
r_1	1.5	2.9			A
r_2	2.5	3.6			A
r_3	3.5	4.2			A
r_c	7.6	7.6			A
k_{ijk} (O-Si-O)	0.5				$eVrad^{-5.2}$
k_{ijk} (Si-O-Si)	0.25				$eVrad^{-8}$
n (O-Si-O)	1.2				
n (Si-O-Si)	4.0				

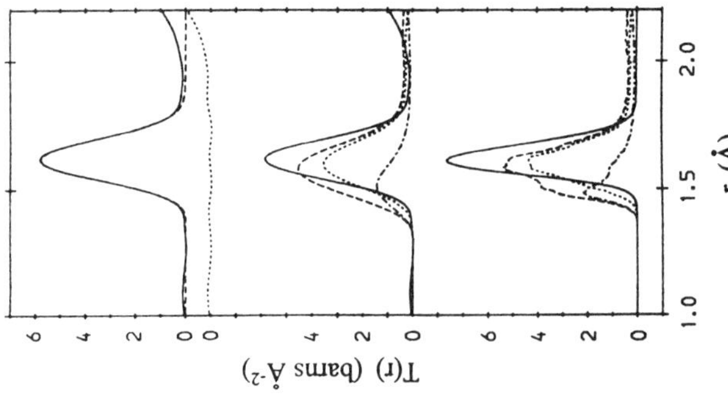

Figure 3 A comparison of the Si-O bond length distribution from the simulation for the Na_2O-SiO_2 glass with the neutron diffraction data.experiment. Top, a fit to the first (Si-O) peak in the neutron correlation function (_____ , experiment; _ _ _ _ , fit and .... , residual); middle, a comparison of the resolution broadened peak for the simulation (_ _ _) with experiment (_____) showing the separate contributions from the non-bridging (_ . _ .) and bridging (........) oxygen atoms and bottom, a comparison of the unbroadened distributions for the simulation with that generated from the neutron peak fit parameters (key as middle).

RESULTS AND DISCUSSION

Figure 1 compares the correlation functions obtained from the molecular dynamics simulations of the three glasses with the neutron diffraction data, after weighting the individual components $t_{ij}(r)$ with the appropriate neutron scattering length products and convolution with the peak function P(r), and the corresponding R_χ factors [3] are given in Table I. The dotted lines in Figure 1 represent the difference between the simulations and experiment and indicate that the major discrepancy occurs in the region of the intratetrahedral Si-O and O-O peaks, mainly due to the fact that these distances in the simulation are slightly shorter than those in the real material. Unlike simulations using 2-body potentials, the width of the intratetrahedral O-O peak is in good agreement with that from the diffraction experiments. However, the bond length distribution is much too broad and at least part of the reason for this can be seen from Figure 3. The top curves show a single distance fit to the first peak in the experimental correlation function for vitreous K_2O-SiO_2, which yields a mean bond length of 1.625 +/- 0.005A with an r.m.s. variation of 0.064 +/- 0.005A. A fit was also performed using separate Si-O distances for the bridging and for the non-bridging oxygen atoms (centre curves) and the resulting unbroadened Si-O peak compared with the same function from the molecular dynamics simulation (bottom curves). It can be seen that the latter is partly resolved into a doublet, with the contribution from the non-bridging oxygen atoms occuring at a lower r to that from the bridging oxygen atoms, whereas the former is a much narrower asymmetric peak. Thus the difference in the Si-O bond length for the bridging and non-bridging atoms is too large in the simulation. The two distance peak fit yields mean bond lengths for the non-bridging and bridging oxygen atoms of 1.596 and 1.639A, respectively, giving a difference of 0.043A. The corresponding quantities for crystalline α-$Na_2O.2SiO_2$ [4] are 1.578, 1.630 and 0.052A, which are within the uncertainties of the fit.

In the experimental correlation function for the soda-silica glass, the Na-O interactions give rise to a well defined shoulder on the low r side of the first O-O peak, but this is much less pronounced in the simulation (c.f. Figure 1) due to the shorter O-O distance and Na-O distances which are conversely slightly too long. Molecular graphics representations of the structures generated by modern simulations of alkali silicate glasses, such as that in Figure 2, strongly suggest that the alkali ions have a non-uniform distribution and hence it is important to try to investigate their local environment. This may be achieved using a modified version of the difference technique [2] as shown by the top two pairs of curves in Figure 4. Interpretation of the simple difference curve [dashed line], obtained by subtracting the correlation function for pure vitreous SiO_2 from that for $xNa_2O.SiO_2$, is complicated by the structure arising from the increase in the intratetrahedral Si-O and O-O distances which occurs on adding Na_2O to SiO_2. It is therefore necessary to work with the modified difference curve [dotted line], generated from the peak fit residuals, as discussed in detail in reference [2]. The latter suggests that the first Na-O distance distribution is asymmetric and this is supported by the Na-O contribution from molecular dynamics simulation which is compared to the modified difference function in Figure 5. A similar asymmetry is seen in the K-O component of the K_2O-SiO_2 simulation and in both M-O distributions for the mixed alkali glass.

The disorder inherent to the vitreous state means that not all of the oxygen atoms in the first neighbor shell around the sodium ions will be non-bridging and, similarly, that the cavities in which some of the Na^+ ions reside will not be sufficiently regular for them to be at the ideal distance from every oxygen atom. In this situation, the positively charged Na^+ ions will tend to adopt the optimum configuration with respect to the negatively charged non-bridging oxygen atoms, rather than the bridging oxygen atoms which only carry the much smaller effective charge associated with heteropolar covalent bonding. Thus it might be expected that the Na-O distribution will comprise of an approximately symmetric peak at the non-bridging oxygen atom distance plus a high r tail arising from the bridging oxygen atoms and that asymmetric peak fit to the modified difference function would reflect mainly the non-bridging oxygen atoms. This is consistent both with the molecular dynamics simulation and with the Na(O) co-ordination number of 3.0 +/- 0.5 extracted from a symmetric peak fit to $\Delta T(r)$, since in an oversized cavity the Na^+ ion will normally only be "in contact" with three oxygen neighbors. The resulting difference correlation function, $\Delta T(r)$, is compared to the Na-O distribution from the MD simulation in Figure 5 (dashed curve), together with the separate contributions for non-bridging and bridging oxygen atoms. The co-ordination number is also lower than would be expected in comparison to crystalline α-$Na_2O.2SiO_2$ [4] where each Na^+ ion has five oxygen neighbors, four in the range 2.290-2.386A [mean Na-O distance 2.347A, c.f. 2.338A from the fit] and the fifth at 2.600A. Interestingly, however, the single non-bridging oxygen atom in the first co-ordination shell of the crystal is not the one at 2.600A but that at 2.386A.

CONCLUSIONS

Although some optimisation of the potentials is still required, the present study clearly demonstrates the powerful role which molecular dynamics can play in interpreting diffraction data for multicomponent glasses. This is particularly true for systems, such as Na_2O-SiO_2, where it is not possible to use techniques such as isotopic substitution to investigate the individual component correlation functions.

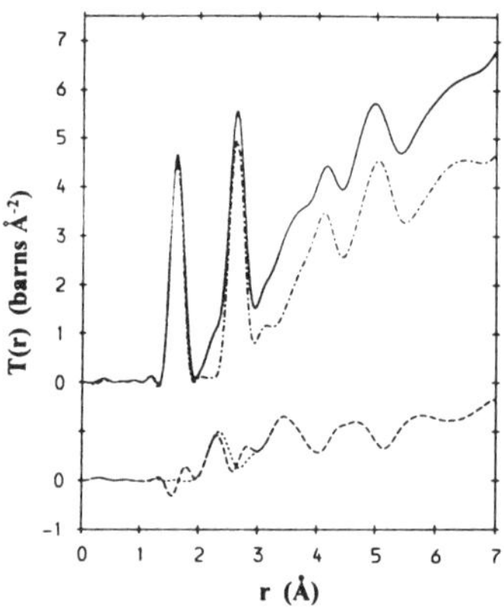

Figure 4 An analysis of the first neighbor Na-O distribution for the Na_2O-SiO_2 glass using the modified difference technique (Wright et al., 1991). From the top: Na_2O-SiO_2 (_________) and SiO_2 (_ . _ . _ . _); difference (- - - - -) and modified difference (. . . . . . .) correlation functions.

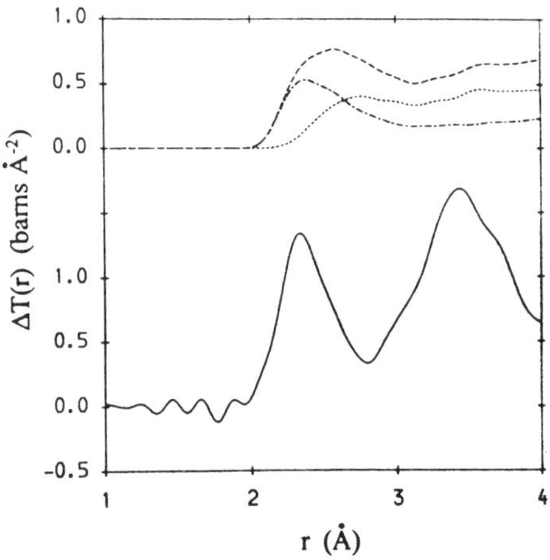

Figure 5 An analysis of the Na-O distribution for the Na_2O-SiO_2 glass using the modified difference technique. Bottom, modified difference correlation function (________) ; top, molecular dynamics simulation showing the total Na-O distribution (- - - -) and the separate contributions from non-bridging (_ . _ . _) and bridging (. . . . . . .) oxygen atoms.

REFERENCES

[1] W.C. LaCourse, A.N. Cormack, "Structural influences on the mixed alkali effect in glasses", *Trans. A.C.A.* **27**, 211-223, 1991.

[2] A.C. Wright, A.G. Clare, B. Bachra, R.N. Sinclair, A.C. Hannon, B. Vessal, "Neutron diffraction studies of silicate glasses", *Trans. A.C.A.* **27**, 239-253, 1991.

[3] A.C. Wright, "The comparison of molecular dynamics simulations with diffraction experiments", *J. Non-Cryst. Solids* **159**, 264, 1993.

[4] A.K. Pant, D.W.J. Cruickshank, "The crystal structure of α-$Na_2Si_2O_5$", *Acta Crystallogr.* **B24**, 13, 1968.

[5] P. Ewald, "Die Berechnung optischer und elektrostatischer Gitterpotentiale", *Ann. Physik.*, **64**, 253, 1921.

[6] M.J.L. Sangster and M. Dixon, "Interionic potentials in alkali halides and their use in simulations of the molten salts", *Advances in Physics*, **25**, 247, 1976.

[7] N. Anastasiou and D. Fincham, "Programs for the dynamic simulation of liquids and solids", *Comput. Phys. Commun.*, **25**, 159, 1982.

TESTING NON-COVALENT GLASS MODELS WITH PAULING'S VALENCE RULE

J.W. Adams, US Army Engineer Waterways Experiment Station, Vicksburg, MS

B. Vessal, Biosym Technologies, Inc., San Diego, CA

B.H.W.S. de Jong, Institute for Earth Sciences and Vening Meinesz Institute for Geodynamic Research, University of Utrecht, Utrecht, the Netherlands

ABSTRACT

A problem with computational models of glasses is that experimental data are lacking against which to explicitly test calculated results. A way to assess the quality of such models is using crystal chemical constraints as embedded in Pauling's valence principle.[1] Crystal chemical analysis of a molecular dynamics (MD) model of lithium disilicate glass shows that the overall charge transfer between cations and anions is well accounted for and that the variation in valence strength donated by silicon and accepted by oxygen falls within proper limits, but that the spread in calculated valence donated by the lithium ions is too large. Thus valence strength calculations provide a tool for iterative atomic potential optimization.

INTRODUCTION

What can be determined from models of glasses? Local environments may be sampled in detail, for example, pair and bond angle distributions, connectivity among atoms, and Q speciation for silica tetrahedra. Medium range topological analyses may also be sampled, requiring more work. But how true are these results vis a vis experimentally observable ones? A problem encountered in model silicate glasses is that many possible non-equilibrium states of a random network of SiO_4 tetrahedra are close in energy to the single energy minimum for the isochemical crystalline form. Models tend to be compared to non-uniquely interpretable NMR, EXAFS, infrared or Raman spectra of glasses. Such lack of uniqueness has for instance been demonstrated in radial distribution analyses of glassy B_2O_3[2] and ^{29}Si NMR silica Q species distribution of lithium silicate glasses.[3]

An alternative route toward model validation rests on using Pauling's valence rule.[1] This rule states that in a stable structure, the total strength of the valency bonds that reach an anion from all neighboring cations in a coordination polyhedron is equal to the total charge of the anion. Its validity is based on the Lewis octet rule which ascribes a nobel gas electronic structure to each atom in a molecule. This rule is still useful for predicting qualitative molecular geometry from the valence shell electron pair repulsion formalism. Pauling did not consider bondlength variations in applying his rule. Valences were fixed, for instance, silicon having a valence of -4, lithium of -1, and oxygen 2. However it was soon recognized that bondlength variations have to be taken into account. The most comprehensive form to date has been parametrized by Brese and O'Keeffe.[4] These authors inspected roughly 1000 atom pairs from well determined crystal structures in order to assign valence V_{ij} donated by atom i and j in a cation-anion linkage according to Equation (1). Here R_{ij} is a tabulated bond valence parameter, and d_{ij} a measured distance between atoms i and j.

$$V_{ij} = \exp\left[\frac{R_{ij} - d_{ij}}{0.37\text{Å}}\right] \qquad (1)$$

This bondlength-dependent parametrized form has proven to account satisfactorily for charge balances in well determined crystalline alkali di- and orthosilicates.[5-8] Its successful application must therefore be considered to provide a demanding test that all crystal chemical requirements which can be put upon possible metric and topologic arrangements of atoms in space have been fulfilled. It is with these strong constraints that we set out to make a detailed assessment of the quality of the results of one molecular dynamics model of glassy lithium disilicate.

EXPERIMENTAL

MD Model Calculation

A constant volume MD model of glassy lithium disilicate, $Li_2Si_2O_5$, was simulated by using state-of-the-art potentials to calculate a glassy structure on the Cray XMP/416 at Rutherford Appleton Laboratory. The structure consisted of 576 atoms in a constant volume monoclinic cell with periodic boundary conditions.[3] Initial atomic coordinates were taken from Liebau's 1961 determination of the crystal structure of $Li_2Si_2O_5$, a layered sheet silicate with a monoclinic unit cell.[9] Pertinent potential parameters used are listed in Table I. Long range coulombic interactions were computed using Ewald's method as outlined by Vessal et al.[10] Buckingham potentials accounted for all other short range atomic pair interactions and were taken from previous MD simulations of silica[10,11] and alkali silicates.[12] Short range 3-body potentials were calculated for O-Si-O and Si-O-Si connections. The equilibrium O-Si-O bond angle was taken as $\theta_O = 109.48°$, whereas for Si-O-Si the equilibrium angle, θ_O, was 144°. Si-O, O-O, Si-O-Si and O-Si-O parameters were taken from potentials fitted to the elastic and dielectric properties of α-quartz.[10,11] The Li-O interatomic pair potentials used were constrained to give equilibrium distances equivalent to those distances given in the crystal.[13] The system was first "melted" in 2500 time steps during 7.5 ps at 6000 K and then cooled for the same number of steps to 2500, 1200, 625 and 293 K. Spatial atomic radial and angular distributions, Q speciation and total correlation functions of the model are discussed elsewhere.[3] Figure 1 shows the pair correlation for this glass.

Table I. Pertinent potential parameters used for the MD $Li_2Si_2O_5$ model. The three-body interactions were calculated using equation (14) of reference 11.

Parameter*	unit	Si-O	O-O	Li-O
A_{ij}	eV	990.6170	4511887.2	292.3
ρ_{ij}	nm	0.03297	0.0149	0.03472
θ_{ijk} ⟨Si-O-Si⟩	eV/rad^2	3779.8		
θ_{ijk} ⟨O-Si-O⟩	eV/rad^2	729.0189		

*A_{ij} and ρ_{ij} are constants and θ_{ijk} is the calculated bond angle in radians

RESULTS

The average valence strengths calculated for the MD model glass are collected in Table 2 together with those of selected crystalline alkali di- and orthosilicates. The R value for the crystalline compounds refers to the quality of the crystal structure determination; the smaller the value, the better the determination.

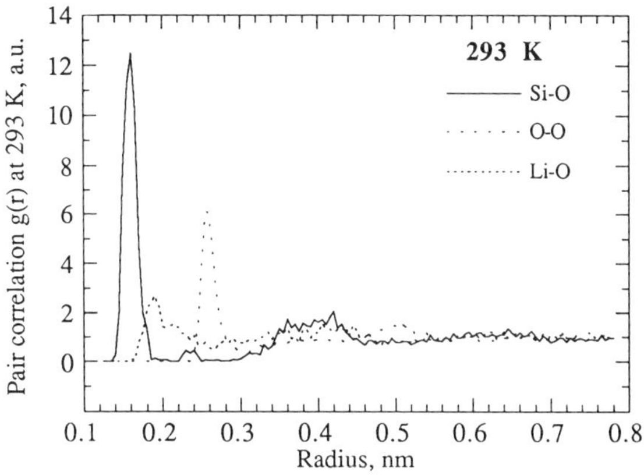

Figure 1. Pair correlation for the MD model lithium disilicate glass.

Table 2. Average valences for crystalline and model silicates.

Compound	R*	⟨Si⟩	⟨Li⟩	⟨O⟩	Ref.
$Li_2Si_2O_5$	0.042	-4.29	-1.02	2.14	5
$Cs_2Si_2O_5$	0.088	-3.98	--	2.09	5
$Cs_{1.33}Li_{0.67}Si_2O_5$	0.0501	-4.12	-1.09	2.06	6
$KLiSi_2O_5$	0.0331	-4.01	-1.01	1.87	7
Li_4SiO_4	0.028	-3.88	-0.97	1.98	8
$Li_2Si_2O_5$ MD glass	--	-4.32	-0.76	2.03	This work

$*R = \left\{\Sigma\left|F_o - |F_c|\right|\right\} / \Sigma F_o$

DISCUSSION

Silicon

The calculated range of silicon valences varies between -3.86 and -4.29 for all crystalline silicates collected in Table 2. The range for tetrahedrally oxygen-coordinated silicon in the MD model of glassy lithium disilicate is similar, varying between -4.04 and -4.60. The crystallographically unique silicon atom in crystalline lithium disilicate has a valence of -4.29.

The average d(Si-O(nbr)) and d(Si-O(br)) in tetrahedrally coordinated silicon in the MD model of glassy lithium disilicate are 0.153 and 0.162 nm respectively. The number of Si-O(br) and Si-O(nbr) linkages is set by the stoichiometry of the sample, being for a glassy disilicate 3 to 1. The average d(Si-O) is 0.1596 nm close to the theoretically required value of 0.1598 nm ([3d(Si-O(br)) + d(Si-O(nbr))] / 4). The corresponding Si valence, -4.32, seems high, presumably due to the unusually small MD derived average d(Si-O(nbr)) vis a vis 0.156 nm in crystalline lithium disilicate.

Oxygen

The calculated range of oxygen valences varies between 1.73 and 2.38 for all crystalline silicates collected in Table 2. The range observed for oxygen in the MD of glassy lithium disilicate varies between 1.6 and 2.4; a variation, just as in the case of the silicon valence, within accepted tolerances.

In crystalline lithium disilicate three different oxygen atoms occur. These are O(br$_I$) connected to two Si atoms, O(br$_{II}$) connected to two Si atoms and one Li atom, and O(nbr) connected to three Li atoms and one Si atom, with valence values of 2.23, 2.17, and 2.03 respectively. For lithium orthosilicate the 21 unique O(nbr) atoms vary in valence between 1.89 and 2.08. Clearly the overall range in oxygen valence for crystalline lithium silicates is substantially narrower than that calculated from our MD model both at the high end as well as at the low end. Thus some oxygen atoms receive too many electrons, others not enough. Considering that the MD model properly calculates the O(br) to O(nbr) ratio as well as ⟨d(O-O)⟩ of 0.255 nm, this variation might be due in part to shortcomings in calculating d(Si-O(nbr)) as noted previously. As discussed below the variation may also be due to shortcomings in calculating d(Li-O) or the lithium coordination polyhedron around each oxygen atom.

Lithium

The lithium valences calculated for all crystalline silicates in Table 2 vary between -0.75 and -1.13, with -1.02 being the specific value of the crystallographically unique lithium atom in lithium disilicate. The lithium valences calculated from our MD model vary between -0.3 and -1.3 valence units. This discrepancy is too large and requires some detailed discussion.

The dependence of V_{ij} on d(Li-O) calculated according to Eq. (1) with $R_{Li\text{-}O}$ of 1.466 is shown in Figure 2. Here is indicated that a lithium atom donates 0.5 valence units to oxygen at d(Li-O) of 0.17 nm, but only 0.08 valence units at a distance of 0.24 nm. The sum of the valence given up by a lithium atom to all oxygen atoms in its coordination sphere should vary at most $^-1 \pm 0.3$ valence units in a properly determined structure. Calculated d(Li-O) corresponding to an ideal lithium valence strength of $^-1.00$ for regular 4, 5, and 6-fold oxygen coordination are 0.198, 0.206 and 0.212 nm respectively as indicated in Figure 2.

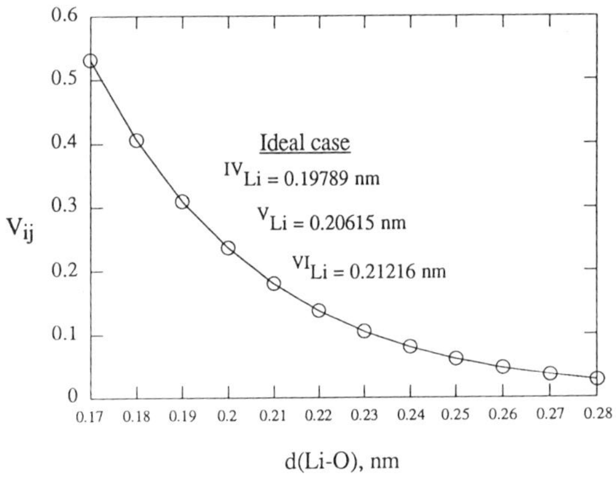

Figure 2. The dependence of V_{ij} on d(Li-O). To donate -1 valence unit to its surrounding oxygens d(Li-O) has to be 0.19789 nm if lithium is four-fold oxygen coordinated, 0.20615 nm if five-fold, and 0.21216 nm if six-fold oxygen coordinated.

We considered any oxygen atom within a d(Li-O) of 0.28 nm as belonging within the lithium coordination sphere, a bondlength which is 15% larger than the longest known d(Li-O) in crystalline silicates.[3,14]. The coordinations collected in Table III are calculated using this cutoff. Calculated lithium valences, i.e. electrons donated by lithium, for the 128 lithium atoms in the MD model of glassy lithium disilicate are given in Table IV.

Table III. Lithium coordination from MD model of glassy lithium disilicate.

Li Coordination number, %	
II	2
III	36
IV	41
V	17
VI	4
Average coordination	3.85

Table IV. Percentage of Li atoms with varying valence according to MD model of lithium disilicate glass.

Valence strength units										
≤-0.3	≤-0.4	≤-0.5	≤-0.6	≤-0.7	≤-0.8	≤-0.9	≤-1.0	≤-1.1	≤-1.2	≤-1.3
1	3	7	15	21	24	15	7	3	3	1

Inspection of Table III shows that the number of 2- and 3-, as well as 5- and 6-fold oxygen coordinated lithium atoms, is unexpectedly high in view of the strictly 4-fold oxygen coordination in crystalline lithium disilicate and the 4-, 5- and 6-fold oxygen coordination in crystalline lithium orthosilicate. What is surprising is the wide range of coordination numbers suggesting a large variety of lithium-oxygen polyhedra, which is not in accordance with Pauling's parsimony rule. The average lithium-oxygen coordination of 3.85 on the other hand seems fine. That all is not well with the lithium potential is more vividly demonstrated by Table IV. Here 47 % of all lithium atoms donate less than 0.7 valence units to the surrounding oxygen atoms. This is not in accordance with acceptable valence strength criteria for stable structures, and in all likelihood also not in pseudostable glassy structures.

CONCLUSIONS

Valence strength criteria enable critical testing of bond features in atomic structures which are experimentally difficult to sample. In the example presented here we show that MD-derived glassy structures can globally do very well indeed in assessing bridging to non-bridging oxygen ratios, in ensuring that each oxygen atom in a structure is bound to at least one silicon atom, and in predicting reasonable geometries of the silica framework. On the other hand, we show that crystallographic awareness as given by valence parametrization models enables unequivocal rejection of structural details, in particular the local environment relating to the non-framework atoms. Although the overall topology of a structure may appear to be correct, a summation of valence values can in fact show that there are some unlikely connections among individual atoms. Thus valence calculations should prove an invaluable tool for improved editing of bond potentials and for assessing the canonical atomic configuration of more complicated structures, e.g., ternary glasses, relative to all possible structures.

ACKNOWLEDGMENTS

BV made the MD models while at Keele University, Staffordshire, England. JWA's work was supported in part by an Ernest Oppenheimer Fund for Surface Science and Colloid Research Fellowship, Cambridge University and by the U.S. Army Environmental Quality Technology Program.

REFERENCES

1. L. Pauling, "The principles determining the structure of complex ionic crystals," *J. Am. Chem. Soc.* **51**, 1010-1026, 1929.

2. M.P. Teter, "Glass structure using quantum molecular dynamics," *Bull. Am. Cer. Soc.*, **66**, 3, 16-G-87, 1987.

3. J.W. Adams, "Morphogenesis in lithium disilicate glass: Internal nucleation in amorphous structures," Ph.D. thesis, Cambridge University, 1993.

4. N.E. Brese and M. O'Keeffe, "Bond-valence parameters for solids," *Acta Cryst.* **B47,** 192-197, 1991.

5. B.H.W.S. de Jong, P.G.G. Slaats, H.T.J. Supèr, N. Veldman and A.L. Spek, "Extended structures in crystalline phyllosilicates: silica ring systems in lithium, rubidium, cesium and cesium/lithium phyllosilicate," *J. Non-Cryst. Solids,* **176**, 164-171, 1994.

6. N. Veldman, A.L. Spek, H.T.J. Supèr, and B.H.W.S. de Jong, "Crystal structure of $Cs_{1.33}Li_{0.67}Si_2O_5$," *Acta Cryst* **C**, in press.

7. N. Veldman, A.L. Spek, H.T.J. Supèr, and B.H.W.S. de Jong, "Crystal structure of $KLiSi_2O_5$," in preparation.

8. B.H.W.S. de Jong, D. Ellerbroek and A.L. Spek, "Low-temperature structure of lithium nesosilicate, Li_4SiO_4, and its Li1s and O1s X-ray photoelectron spectrum," *Acta Cryst.*, **B50**, 511-518, 1994.

9. F. Liebau, "Untersuchungen an Schichtsilikatendes Formeltyps $A_m(Si_2O_5)_n$. I. Die Kristallstruktur der Zimmertemperaturform des $Li_2Si_2O_5$," *Acta Cryst.*, **14**, 389-395, 1961.

10. B. Vessal, M. Leslie and C.R.A. Catlow, "Molecular dynamics simulation of silica glass," *Molec. Simul.* **3**, 123-136, 1989.

11. B. Vessal, M. Amini, D. Fincham and C.R.A. Catlow, "Water-like melting behaviour of SiO_2 investigated by the molecular dynamics simulation technique," *Phil. Mag. B* **60**, 753-755, 1989.

12. B. Vessal, M. Leslie and C.R.A. Catlow, "Molecular dynamics simulation of alkali silicate glasses," *MD and MC Simulation* **28**, 39-47, 1988.

13. S.C. Parker, Ph.D. thesis, University College of London, 1982.

14. R.D. Shannon, "Revised effective ionic radii and systematic studies of interatomic distances in halides and chalcogenides," *Acta Cryst.* **A32**, 751-767, 1976.

MILLION ATOM SIMULATIONS OF MATERIALS ON PARALLEL COMPUTERS - SILICA, SILICON NITRIDE, AND CERAMIC FILMS

Aiichiro Nakano, Rajiv K. Kalia, and Priya Vashishta
Concurrent Computing Laboratory for Materials Simulations
Department of Physics and Astronomy
Department of Computer Science
Louisiana State University, Baton Rouge, LA 70803

ABSTRACT

Multiresolution molecular dynamics approach on parallel computers has been used to investigate the structural properties and mechanical failure in microporous silica. Critical behavior at fracture is analyzed in terms of pore percolation and kinetic roughening of fracture surfaces. Crack propagation in amorphous silicon nitride films is investigated, and a correlation between the speed of crack propagation and the morphology of fracture surfaces is observed.

INTRODUCTION

Porous silica has been the focus of many investigations [1]. This environmentally safe material has numerous technological applications: It is used in thermal insulation of commercial and household refrigerators; in passive solar energy collection devices; in particle detectors; and in catalysis and chemical separation. There is an exciting possibility of utilizing it as an embedding framework in optical switches made of quantum-confined microclusters. Since these applications are due to the remarkable porous structure of the system, it is important to understand the size and spatial distributions of pores and the morphology of pore interfaces.

Silicon nitride has been at the forefront of research for high-temperature, high-strength materials owing to its outstanding properties [2]. The combination of low thermal expansion and high strength makes silicon nitride one of the most thermal-shock-resistant materials currently available. The strong covalent bonding between the atoms results in a superb resistance of mechanical deformation and chemical corrosion. Additionally, silicon nitride based films have found a number of applications in microelectronics technology. For these applications, the mechanism of fracture is one of the most important issues.

The morphology of fracture surfaces has drawn a great deal of attention in recent years [3-7]. It is now well-established that a fracture surface, z(x,y), is a self-affine object in that it remains invariant under the transformation, $(x, y, z) \to (ax, ay, a^{\zeta}z)$, where ζ is known as the roughness exponent. A decade ago, Mandelbrot and coworkers reported the first measurements of the roughness exponent for metallic surfaces [3]. Since then there has been a great deal of controversy regarding the value of ζ. Bouchaud, Lapasset, and Planés carried out measurements of ζ for four aluminum alloys with different heat treatments and in each case they obtained $\zeta = 0.8$ [4]. Måløy et al. made measurements on six different brittle materials and found ζ to be 0.87 ± 0.07 [5]. This led them to conjecture that fracture surfaces of brittle and ductile materials had a "universal" roughness exponent, independent of material characteristics and the mode of fracture. Milman et al. questioned the validity of the "universality" of ζ, especially at microscopic length scales, by pointing out that their scanning tunneling microscopy data for MgO, Si, and Cu revealed the roughness exponent to be around 0.6 [6]. Measurements on tungsten and graphite also indicated a low value of ζ (≈ 0.4) [6]. However, recent molecular-dynamics (MD) simulations for such disparate systems as porous silica [7] and two-dimensional Lennard-Jonesium [8] found $\zeta \approx 0.8$.

In this paper we report the results of a MD study of (i) roughening of fracture surfaces in microporous silica and (ii) crack propagation and the morphology of fracture surfaces in amorphous silicon nitride films.

MULTIRESOLUTION MOLECULAR DYNAMICS ALGORITHM

Molecular dynamics (MD) approach provides the phase-space trajectories of particles through the solution of Newton's equations. The compute-intensive part of MD simulations is the calculation of interparticle interactions. We are dealing with materials in which interatomic interactions are characterized by steric repulsion, Coulomb and charge-dipole interactions, and three-body covalent interactions. Highly efficient algorithms have been designed to compute these interactions on parallel machines [9]. The long-range Coulomb interaction is calculated with a divide-and-conquer scheme, called the fast multipole method

(FMM), which reduces the computational complexity from $O(N^2)$ to O(N). For short-ranged two- and three-body interactions, we have employed a multiple time-step (MTS) approach in which the force on a particle is subdivided into primary, secondary, and tertiary components (see Fig. 1). A significant reduction in computation is achieved by exploiting different time scales of these force components.

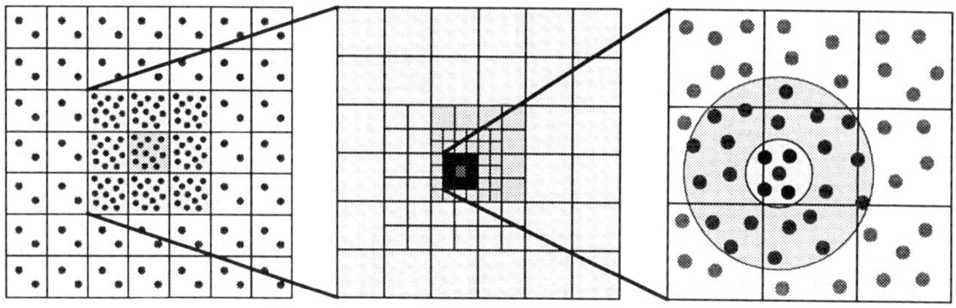

FIGURE 1: Schematic representation of the multiresolution algorithm. (a) Periodically repeated images of the original MD box. Replacing each well-separated image by a small number of particles with the same leading multipole expansions reduces the computation enormously while maintaining the necessary accuracy. (b) A hierarchy of cells in the fast multipole method. (c) The near-field forces on a particle are due to primary, secondary, and tertiary neighbor atoms.

Figure 2 shows the performance of this approach on the Touchstone Delta and the IBM SP1 machine at Argonne National Laboratory. For a 4.2 million particle silica glass, the execution time for a single MD time step is only 4.84 seconds on the 512-node Delta machine. The execution time scales linearly with the size of the system and the computation dominates the communication time.

FRACTURE OF SILICA GLASS

Structural properties and mechanical failure in porous silica have been investigated with million-particle MD simulations [7]. The results for pore size distribution, internal surface area and surface-to-volume ratio of pores, and fractal dimension are in accordance with structural measurements. We have performed

MD calculations on 1.12-million particle amorphous silica systems, investigating the growth of pores with a decrease in the density of the system. As the normal-density glass is uniformly expanded, the pores begin to form when the density of the system is reduced to 1.8 g/cm^3. Further decrease in the density of the system causes an increase in the number of pores and also the pores coalesce to form larger entities (see Figure 3). There is a dramatic increase in the size of pores when the mass density is reduced to the critical value, ρ_c = 1.4 g/cm^3. At the critical density, some pores percolate through the entire system causing fracture. In Fig. 4 we show one of the surfaces of the percolating pore.

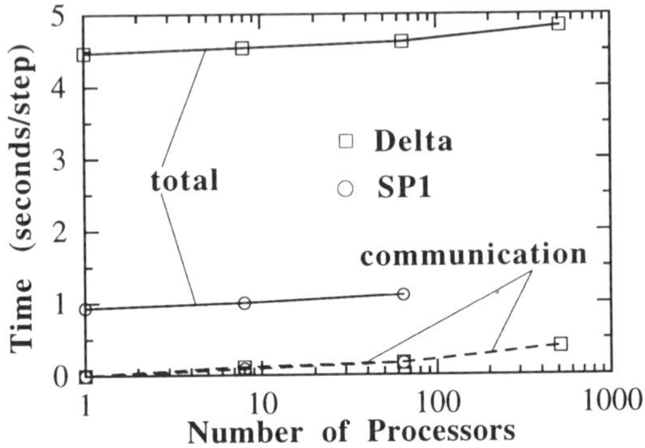

FIGURE 2: Execution time per step for multiresolution MD simulations of amorphous SiO_2 (solid curves) with 8,232p particles where p is the number of processors. Open squares and open circles are the results on the Delta and SP1, respectively. Dashed curves denote the communication overhead.

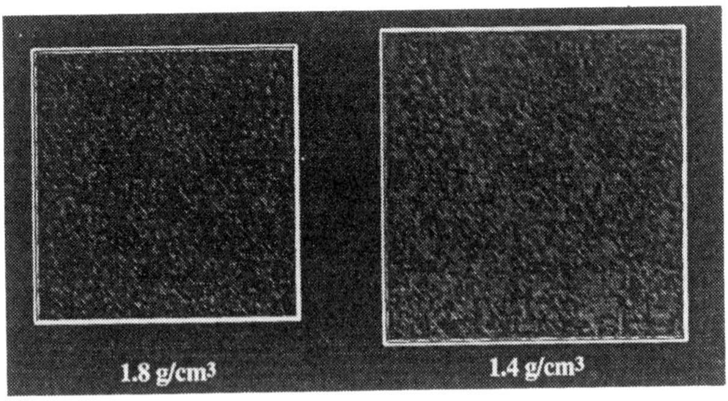

FIGURE 3: Snapshots of two-dimensional slices of MD configurations of silica (1.12 million particles) at densities 1.8 g/cm^3 and 1.4 g/cm^3. Red and blue colors represent pores and silica, respectively.

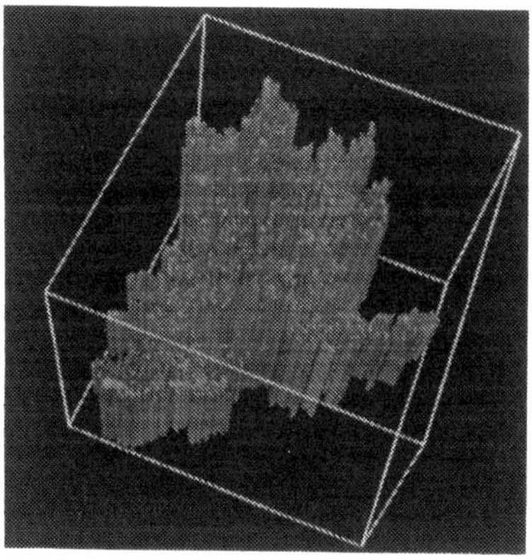

FIGURE 4: Snapshot of a fracture surface resulting from a percolating pore in silica glass at a mass density of 1.4 g/cm^3.

The roughness of a fracture surface is calculated from the height-height correlation function, $G(\sigma)$, where σ is the horizontal distance. The MD results for $G(\sigma)$ are well-described by the relation, $G(\sigma) \sim \sigma^{\zeta}$ with $\zeta = 0.87 \pm 0.02$ for $\sigma < 10$ nm (see Fig. 5). The MD results for the roughness exponent agree with experimental measurements, thus lending further support to experimental claims that the roughness exponent of fracture surfaces is a material-independent quantity. Moreover, the MD results indicate that the universality of the roughness exponent may prevail even at length scales ≤ 10 nm.

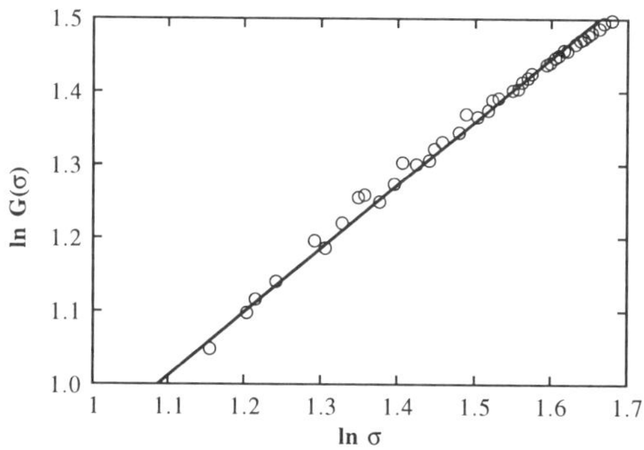

FIGURE 5: Height-height correlation function (open circles) versus the in-plane distance, σ, for the fracture surface shown in Fig. 4. The solid line is the best fit, $G(\sigma) \sim \sigma^{\zeta}$ with $\zeta = 0.87 \pm 0.02$ for $\sigma < 10$ nm

CRACK PROPAGATION IN SILICON NITRIDE FILMS

Molecular-dynamics study of fracture in amorphous Si_3N_4 films involved systems with 100,352 atoms (typical dimensions of a film were 220Å × 220Å × 20Å), interacting via a combination of two- and three-body potentials. The two-body potential includes steric repulsion, the effect of charge-transfer via Coulomb interaction, and the large electronic polarizability of anions through the charge-dipole interaction [10]. Three-body interactions with bond-bending and bond-stretching terms were introduced to include covalent effects [10].

We first prepared well-thermalized bulk system by quenching the molten state and then periodic boundary conditions were removed and the systems were relaxed with MD and conjugate-gradient methods. These well-thermalized crystalline and amorphous Si_3N_4 films were subjected to uniaxial tensile loads by displacing atoms uniformly in the uppermost and lowermost layers (thickness ~ 5.5Å each) along the x direction. The strain was applied at a constant rate while maintaining the temperature at 300 K.

To investigate crack propagation, we insert a crack in an uniaxially stretched film (strain ~ 4 %) by removing particles within a region whose projection onto the xy plane is 4 Å × 50 Å, and we use a strain rate of 0.01 ps^{-1}. The crack plane is parallel to the yz plane, and the crack propagates along the y direction.

Figure 6 shows snapshots of the amorphous Si_3N_4 film projected onto the xy plane. Initially the crack propagates straight along the y direction, as shown in Fig. 6 (a), for the first 4.5 ps. At 7.9 ps, we observe the formation of voids in front of the crack tip, see Fig. 6 (b). These voids grow and form a secondary crack at t = 10.8 ps, see Fig. 6 (c). Eventually the secondary crack and the initial crack coalesce and the resulting crack surface is very rough, as evident from the snapshot in Fig. 6 (d) at t = 16.4 ps.

We have calculated the height-height correlation function [7],

$$g(y) = \left\langle [h(y+y_0) - h(y_0)]^2 \right\rangle^{1/2},$$

of the crack surface using the height profile, h(y). In the equation, the bracket denotes the average over y_0. For a self-affine surface, the height-height correlation function is expected to obey the scaling relation, $g(y) \sim y^{\zeta}$ [3-7].

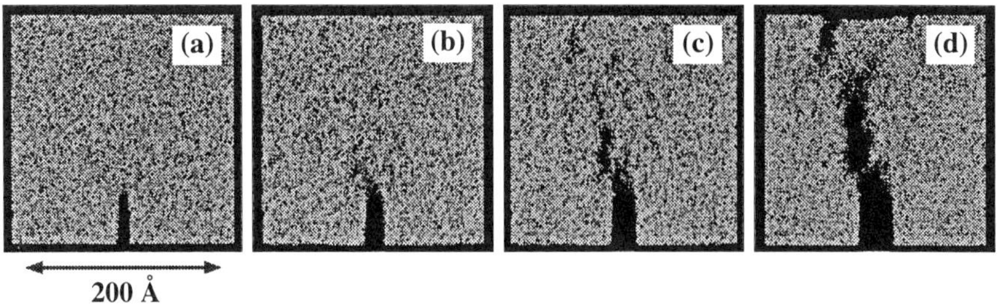

FIGURE 6: Snapshots of crack propagation in an amorphous Si_3N_4 film at time (a) 4.5 ps, (b) 7.9 ps, (c) 10.8 ps, and (d) 16.4 ps.

Figure 7 shows a log-log plot of the height-height correlation function for the crack surface in the amorphous film. There are two well-delineated regimes in this figure. For smaller length scales, we observe a rather smooth crack surface with a smaller roughness exponent. By linear fitting, we obtain a roughness exponent of 0.44 ± 0.02 for $y < \xi = 25$ Å. Beyond this crossover length, ξ, the surface is rougher with a larger exponent, $\zeta = 0.82 \pm 0.02$.

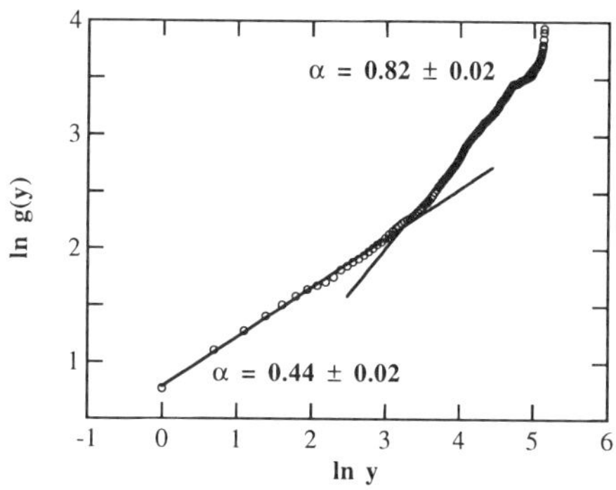

FIGURE 7: Log-log plot of the height-height correlation function, g(y) (open circles). The solid curves represent the best fit, $g(y) \sim y^{\zeta}$, with $\zeta = 0.44 \pm 0.02$ for $y < \xi = 25$ Å, and $\zeta = 0.82 \pm 0.02$ for $y > \xi$.

To relate the morphology of the crack surface to the crack dynamics, we show in Fig. 8 the crack tip position as a function of time in the amorphous film. Initially the crack propagates slowly and continuously. At 12 ps the crack tip jumps suddenly as the initial and secondary cracks coalesce. The distances between this and each of the subsequent jumps are between 20 and 40 Å, which are close to the crossover length of the height-height correlation function. We thus conclude that the smaller roughness exponent (ζ = 0.44) corresponds to quasi-static crack propagation inside microcracks, and the larger exponent (ζ = 0.82) corresponds to the inter-microcrack propagation associated with the coalescence of microcracks. The average propagation speed in the second regime (1630 m/s) is much larger than in the first regime (640 m/s).

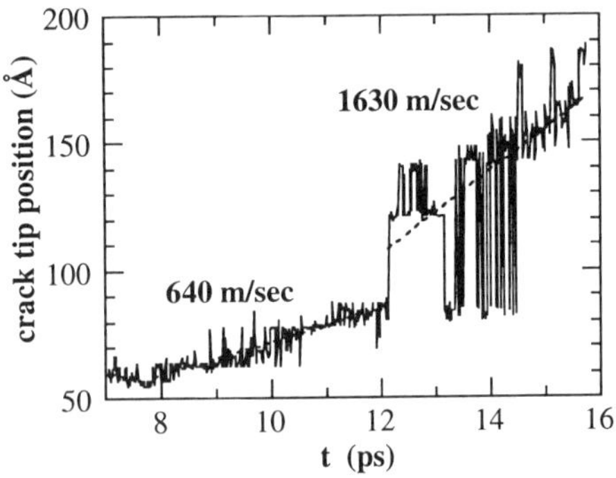

FIGURE 8: Crack tip position as a function of time (solid curves) in an amorphous Si_3N_4 film. Linear fits (dashed lines) before and after 12 ps give the average crack tip velocities of 640 m/s and 1630 m/s, respectively.

The crossover from quasi-static to rapid fracture was also observed in recent experiments by Bouchaud and Navéos [11]. For titanium aluminum alloys, they observed a crossover of the roughness exponent from 0.45 to 0.84. They also found that the crossover length scale decreases when the local stress intensity factor, or correlatively, the crack velocity increases. In MD simulations, the very large tensile stress results in a small crossover length on the nanometer scale.

This crossover phenomenon is consistent with a model in which a fracture surface is considered the trace of a line propagating in a random medium [11,12]. The model predicts the roughness exponent to be 0.75 for high-velocity and 0.5 for low-velocity fracture surfaces [13].

This work was supported by the U.S. Department of Energy, Grant No. DE-FG05-92ER45477, National Science Foundation, Grant No. DMR-9412965, and Air Force Office of Scientific Research, Grant No. F 49620-94-1-0444. A part of these simulations were performed on the 128-node IBM SP computer at Argonne National Laboratory. The computations were also performed on parallel machines in the Concurrent Computing Laboratory for Materials Simulations (CCLMS) at Louisiana State University. The facilities in the CCLMS were acquired with the Equipment Enhancement Grants awarded by the Louisiana Board of Regents through Louisiana Education Quality Support Fund (LEQSF).

REFERENCES

1. J. Fricke, J. Non-Cryst. Solids **121**, 188 (1990).
2. J. Mukerji, in *Chemistry of Advanced Materials*, ed. C. N. R. Rao, Blackwell (Oxford, 1993).
3. B. B. Mandelbrot, D. E. Passoja, and A. J. Paullay, Nature **308**, 721 (1984).
4. E. Bouchaud, G. Lapasset, and J. Planès, Europhys. Lett. **13**, 73 (1990).
5. K. J. Måløy, A. Hansen, E. L. Hinrichsen, and S. Roux, Phys. Rev. Lett. **68**, 213 (1992); ibid. **71**, 205 (1993).
6. V. Y. Milman, R. Blumenfeld, N. A. Stelmashenko, and R. C. Ball, Phys. Rev. Lett. **71**, 204 (1993).
7. A. Nakano, R. K. Kalia, and P. Vashishta, Phys. Rev. Lett. **73**, 2336 (1994). For fracture in ceramic materials, see Refs. 14 and 15.
8. F. F. Abraham, D. Brodbeck, R. A. Rafey, and W. E. Rudge, Phys. Rev. Lett. **73**, 272 (1994).
9. R. K. Kalia, S. W. de Leeuw, A. Nakano, D. L. Greenwell, and P. Vashishta, Comput. Phys. Commun. **74**, 316 (1993); A. Nakano, P. Vashishta, and R. K. Kalia, *ibid*. **77**, 302 (1993); A. Nakano, R. K. Kalia, and P. Vashishta, *ibid*. **83**, 197 (1994).
10. P. Vashishta, R. K. Kalia, and I. Ebbsjö, to be published.
11. E. Bouchaud and S. Navéos, preprint.
12. J. P. Bouchard, E. Bouchard, G. Lapasset, and J. Planès, Phys. Rev. Lett. **71**, 2240 (1993)
13. D. Etras and M. Karder, Phys. Rev. Lett. **69**, 889 (1992).
14. R. Ochoa, T. P. Swiler, and J. H. Simmons, J. Non-Cryst. Solids **128**, 57 (1991); T. P. Swiler, J. H. Simmons, and A. C. Wright, *ibid.* **182**, 68 (1995).
15. A. Sawamura, Y. Watanabe, and R. Yamamoto, *Computer Aided Innovation of New Materials II*, eds. M. Doyama et al. (Elsevier, Amsterdam, 1993).

AN INFRARED SPECTROSCOPY ANALYSIS OF MgO-DOPED SILICON NITRIDE

F. Mohammadi, J. H. Adair, E. D. Whitney, and J.K. West
Department of Materials Science and Engineering
University of Florida
Gainesville, FL 32611

ABSTRACT

Silicon nitride hybrid ball bearings used in high temperature applications undergo mechanical and environmental degradation. To study the surface chemistry of silicon nitride, a CAChe™ Worksystem* has been used to generate the clusters and corresponding transmission vibrational spectra of silicon nitride. In the present study, the effect of surface conditions on the surface chemistry and wear degradation of silicon nitride was evaluated. Infrared reflection spectroscopy (IRRS) used to determine molecular orientations shows a difference in reflectance spectra for fractured and as-received polished samples.

INTRODUCTION

Silicon nitride (Si_3N_4)-based ceramics are the most promising materials for turbine engines and ball bearing applications.[1,2] The properties of sintered silicon nitride (SSN), including resistance to thermal shock, corrosion, creep, oxidation, and wear, has generated interest for applications requiring high temperature structural materials.[3,4]

The preparation and fabrication of high purity silicon nitride products requires consolidation of silicon nitride powders at high temperatures and pressures. The expense of such a process limits the number of economically feasible applications of this material. In addition, because of the inherent low sinterability of Si_3N_4, sintering aids such as Y_2O_3, MgO, and Al_2O_3 are used to decrease the sintering temperature. Therefore, the final sintered silicon nitride (SSN) consists of β-Si_3N_4 grains and a complex grain boundary phase formed by the sintering aids. The grain boundary phases affect both ambient and high temperature properties of SSN.[5,6]

The objective of this work is development of a model for determining the wear mechanisms and lifetime prediction in hybrid Si_3N_4-M50 steel ball bearing systems. Since in jet engine applications, silicon nitride ball bearings are exposed to oxygen and water vapor, present in the atmosphere at high temperatures, the oxidation of silicon nitride under these atmospheric conditions might cause wear degradation and failure under Hertzian contact pressure of Si_3N_4 ball bearings. Thus a study of the oxidation and surface chemistry of Si_3N_4 may yield a better understanding of high temperature surface degradation of silicon nitride.

*CAChe is a registered trademark of CAChe Scientific, Oxford Molecular Group, Beaverton, OR

EXPERIMENTAL PROCEDURES

Determination of molecular conformation and stereochemistry was performed using an infrared spectrometer (Nicolet 20SXB) on as-polished and the fracture surface of a commercial silicon nitride bearing ball. A 0.9525 cm (3/8 in.), as-received, commercial silicon nitride ball was used for both the as-polished and the fracture surface for IRRS measurements.

Generation of the related cluster and calculation of corresponding transmission IR spectra were performed using the AM1 molecular orbital theory based on the quantum mechanical method in a CAChe™ Worksystem. Since the transmission IR spectra of a fracture surface cannot be measured by infrared spectrometers, the experimental reflection spectra were converted to the transmission spectra by inversion to aid in the comparison to the theoretical transmission spectra generated by AM1.

MOLECULAR ORBITAL (MO) METHODS

Calculational chemistry is basically the study of the interactions of atoms by solving Schrödinger's Equation, $H\hat{\psi} = E\hat{\psi}$ for the electrons in the system of atoms being studied.[7] The field of calculational chemistry is broken up into three main areas: statistical mechanics, quantum chemistry, and molecular dynamics.[8] Quantum chemistry is generally split into *ab-initio* (rigorous) and semi-empirical (simplified) methods.[9]

Ab-initio methods treat all of the electrons in an assembly of atoms whereas semi-empirical methods consider only the valence electrons. The computational (CPU) time increases for *ab-initio* methods (STO-3G, 3-21G, 6-31G*, 6-31G**, MPZ, MP4) as well as for cluster size. The semi-empirical methods,[10-12] AM1 and PM3 give very good results with only a modest increase in calculation (CPU) time.[13-15] The errors increase dramatically for the Hückel methods but a very large collection of atoms may be modeled.

In contrast, ZINDO[16] (Zerner's Intermediate Neglect of Differential Overlap program) can use *d* orbitals and be used for many transition series metals. ZINDO also has an alternative parameterization for electronic (UV/visible) spectra which produces results consistent with experiment. Since MOPAC is parameterized for ground-state heats of formation, excitation energies determined by MOPAC will probably not be as good as those from ZINDO. Although the ZINDO geometry optimization is faster than MOPAC, the geometries obtained are not as reliable as those from most MOPAC models (MNDO, AM1, and PM3), but ZINDO geometries are nearly as accurate as the MOPAC MINDO/3 geometries.[17]

As a calculational strategy, the geometries for silicon nitride structures used in this study were optimized using AM1. These optimized geometries were then submitted to a Force calculation within MOPAC to predict the vibrational spectra for the two different ring structures: a small cluster, $Si_3N_4H_8$, and a larger cluster, $Si_8N_{15}H_{16}$, as shown in Figures 1 and 2.

RESULTS AND DISCUSSION

INFRARED SPECTROSCOPY (IR)

Infrared (IR) spectrometry, one of the most useful techniques for obtaining information about molecular structures, has the advantage of providing nondestructive analysis under atmospheric pressure.[18] FTIR spectroscopy was used to analyze both an as-polished surface and a fracture surface of a commercial silicon nitride ball. The reflectance in the IR spectra for the fracture surface was obtained at various regions from the edge to the center of specimen. The IR spectra reveal that β-Si_3N_4 is the predominant phase in the silicon nitride ball bearings, a finding consistent with Tsuge.[19]

IR spectra of as-received silicon nitride samples are shown in Figure 1. Based on earlier reports, two Si-N-Si molecular stretching and three Si-N-Si molecular rocking vibrations are characteristic of Si_3N_4 materials.[20-23] In this study, these surface modes have been determined, as shown in Table I. Molecular vibrations for the wavenumbers above 1500 cm^{-1} were not detected by IRRS, but a peak at 3600 cm^{-1} has been detected corresponding to surface N-H absorption mode. The IR spectra of the fracture surface were almost the same on different regions of the surface, but instead of two peaks at 1070 and 990 cm^{-1} for the as-received, polished specimens, four sharp peaks were observed at 1115, 1040, 950, and 900 cm^{-1} on the fracture surface.

SEM photomicrographs show that the predominant mode of fracture in silicon nitride is intergranular which exposes the sintering aids in the grain boundaries. Since more grain boundary surface area is exposed to the IR beam than that of as-polished surface, a shift in peaks in IR spectra of fracture surface is expected.

CALCULATED SPECTRA

The purpose of these calculations was to develop a model for failure prediction of silicon nitride ball bearings in extreme atmospheric conditions, when they are exposed to oil, oxygen, and water vapor present in air. Calculations have been performed in a Macintosh environment, using a CAChe™ Worksystem (Beaverton, OR).

The theoretical and experimental infrared spectra for the Si_3N_4 clusters are summarized in Table I.

IR spectra containing five peaks, characteristic of silicon nitride materials, are shown in Figures 1 and 2. Figure 1 shows the smallest Si_3N_4 cluster with two rings with hydrogen atoms as the termination atoms on the cluster surface and the corresponding theoretical spectra as well as the experimental vibrational spectra from the fracture surface of a commercial silicon nitride ball.

The theoretical vibrational spectra of the small cluster is in good agreement with the IR spectra of the fracture surface, as shown in Figure 1. In addition, both spectra have similar sharper peaks than that of the as-polished surface. One of the possible explanations for the broader peaks is the existence of amorphous regions at the grain boundaries, composed of small clusters of Si_3N_4 interspersed with the sintering aid materials.

A large cluster of silicon nitride, representative of the bulk material, and the corresponding theoretical and experimental IR spectra are shown in Figure 2. The theoretical peaks become broader and shift to higher wavenumbers similar to the as-polished IR spectra, indicating that the theoretical cluster generated by AM1 corresponds to the bulk material. Small differences between the theoretical and experimental spectra might be due to the presence of sintering aids in sintered silicon nitride ball bearings.

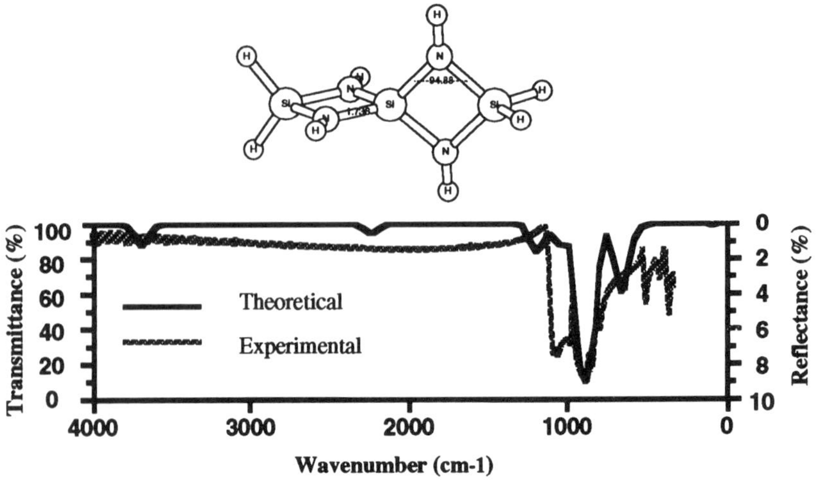

Figure 1. A small silicon nitride cluster, corresponding theoretical transmission and experimental reflection spectra

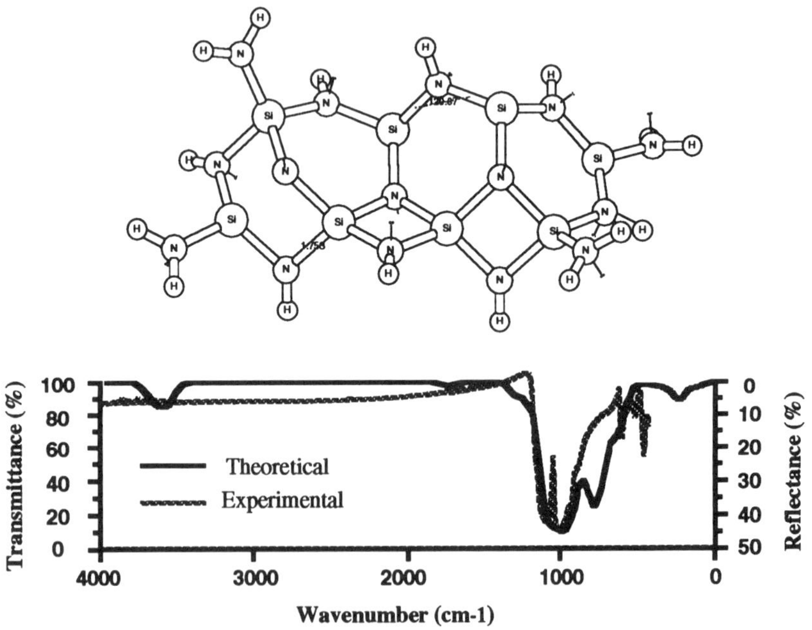

Figure 2. A large silicon nitride cluster, corresponding theoretical transmission and experimental reflection spectra

Table I. Comparison of theoretical transmission and experimental reflection IR of silicon nitride

Theoretical $Si_8N_{15}O_{16}$	Theoretical $Si_3N_4H_8$	Vibrational Modes	Experimental As-Polished	Experimental Fracture	Reference
200	100	N-H rocking	430	430	20,23
580	658	N-H stretching	490	490	20,23
770	886	Si-N-Si stretching	570	570	20,23
980	900	Si-N-Si rocking	835	850	20,23
1040	1040	Si-N-Si stretching	920	900	24,25
1330	1190	N-H stretching	990	950	24,26
1730	2247	N-H stretching	1070	1040	24,26
3600	3646	N-H stretching	1100	1115	24,26

CONCLUSION

Infrared spectroscopy data show reasonable agreement between the AM1-generated and experimental IR spectra. Based on comparisons of theoretical and experimental spectra, large clusters are more characteristic of the as-polished Si_3N_4-rich surface. On the other hand, small clusters are characteristic of the amorphous regions found in grain boundaries of the fracture surface.

Sintering aids are concentrated at the grain boundaries along with silicon nitride clusters. In future work, the addition of spectra from Si-O-N-H, Al-O-N-H, and MgO-N-H systems will be investigated. In addition, ZINDO method for mechanisms of bond formation and AM1 for reaction pathway calculations will be used in future studies. Theoretical modeling followed by experimental characterization using atomic force microscopy, high resolution electron microscopy, and FT-IR on wear debris and oil lubricants will be used for lifetime prediction in the silicon nitride-M50 steel hybrid ball bearing system.

ACKNOWLEDGMENT

The authors wish to thank the Air Force Office of Scientific Research for partial support under grant number F49620-93-1-0349DEF.

REFERENCES

1. L.D. Wedeven and T.A. Harris, "Rolling Element," *Machine Design*, **August 6, 1987**, 72-76.
2. S.T. Buljan and V.K. Sarin, "Silicon Nitride-based Composites," *Composites*, **18** [2], 99-107 (1987).
3. C. Adelhelm and D. Hirschfeld, "Inter-laboratory Comparison of the Chemical Analysis of Silicon Nitride," *Fresenius's Journal of Analytical Chemistry*, **342**, 125-127 (1992).
4. P.E.D. Morgan, "Structuring Chemical Technology to Produce Cost-Effective Ceramic Products on the Large Scale," *Am. Ceram. Soc. Bull.*, **72** [7], 65-70 (1993).

5. H. Sakai, T. Soma, M. Matsui, and I. Oda, "Oxidation and Microstructure of Sintered Silicon Nitride," *Ceramic Microstructures*, **1986**, 137-144.
6. G.T. Burns and G. Chandra, "Pyrolysis of Preceramic Polymers in Ammonia: Preparation of Silicon Nitride Powders," *J. Am. Ceram. Soc.*, **72** [2], 333-337 (1989).
7. Donald B. Boyd, "Aspects of Molecular Modeling," in *Reviews in Computational Chemistry*, K.B. Lipkowitz and D.B. Boyd, Eds., VCH Publishers, Inc., New York, 321-354, **1990.**
8 R.J. Bartlett, "Chemical Computation Has Come Age," *Florida School on Applied Molecular Orbital Theory*, Gainesville, Florida, **1993**.
9. J.A. Pople and D.L. Beaveridge, "Approximate Molecular Orbital Theory," *McGraw-Hill Book Company*, New York, **1970**.
10. MOPAC Version **6.1**, Tektronix, Inc., CAChe Scientific, Beaverton, OR.
11. M.J.S. Dewar, E.G. Zoebisch, E.F. Healy, and J.P. Stewart, "AM1: A New General Purpose Quantum Mechanical Molecular Model," *J. Am. Chem. Soc.*, **107**, 3902-3909 (1985).
12. M.J.S. Dewar and C. Jie, "Organometallics," *J. Am. Chem. Soc.*, **6**, 1486-1490 (1987).
13. Taipau Chia, Jon K. West, and Larry L. Hench, "A Molecular Orbital Model of Gel-Silica IR Spectra," in *Better Ceramics Through Chemistry* **VI**, Materials Research Society, MRS, John Wiley and Sons, New York, in press.
14. Jon K. West and Larry L. Hench, "A PM3 Molecular Orbital Model of Silica Rings and Their Vibrational Spectra," *J.C.C.S.* , accepted for publication, **1994**.
15. Jon K. West and Larry L. Hench, "Molecular Orbital Models of Silica Rings and Their Vibrational Spectra," *J. Am. Ceram. Soc.*, submitted for publication, **1994**.
16. ZINDO Version **3.0**, CAChe Scientific, Inc., Beaverton, OR.
17. Michael C. Zerner, "Semi-Empirical Molecular Orbital Methods." *Review of Computational Chemistry II*, K.B. Lipkowitz and D.B. Boyd, Eds., VCH Publishing, New York, 313-366, **1991**.
18. C. Marcott, "Infrared Spectroscopy," in *ASM Handbook*, Edited by the Materials Information Society, **10**, 109-125 (1992).
19. A. Tsuge, Y. Uwamino, T. Ishizuka, and K. Suzuki, "Quantitative Analysis of Powdery Sample by Diffuse Reflectance Infrared Fourier Transform Spectrometry: Determination of the β-component in Silicon Nitride," *Applied Spectroscopy*, **45** [8], 1377-1380 (1991).
20. L.L. Hench, P.N. Vaidyanathan, and S. Dutta, "Effect of Y_2O_3 and Al_2O_3 on the Oxidation Resistance of Si_3N_4," NASA Lewis Research Center.
21. G.M. Nikolic, "Structure Investigation of Photo-CVD Silicon Nitride Films by IR and ESR Spectroscopy," *Vacuum*, **40** [1-2], 143-144 (1990).
22. A. Takase and E. Tain, "IR Absorption and Raman Spectra of Sintered Si_3N_4 Under Pressure," *Am. Ceram. Soc. Bull.*, **65** [12], 1597-1600 (1986).
23. L.L. Hench and S.W. Freiman, "Investigation of the Oxidation of Hot-pressed Si_3N_4 with Infrared Reflection Spectroscopy (IRRS)," *J. Mater. Sci.*, **16**, 2767-2773 (1981).
24. J.H. Adair, "Interactions of the Surface of Silicon Nitride Particles with Solvents and Organic Molecules," **October 1990.**
25. I. Tejedor, "Ceramic Powder Characterization Using I-GSC," University of Wisconsin at Madison, May 1989.
26. P. Toro and O. Wittke, "Rheological Behavior of Silicon Nitride Whiskers," in *Silicon Nitride Ceramics*, Edited by I.W. Chen, P.F. Becher, M. Mitomo, G. Petzow, and T.Sh. Yen, **287**, 271-276 (1992).

Surface and Interface Structure

SIMULATION OF SURFACE FORMATION IN SOLIDS BY MOLECULAR DYNAMICS FRACTURE

Joseph H. Simmons, Thomas P. Swiler,* Yueh-Long Tsai,
John J. Mecholsky, Jr. and Rajiv Bendale
University of Florida
Gainesville, FL 32611, USA

* now at Sandia National Laboratories, Albuquerque, NM 87185, USA

Abstract

The Molecular Dynamics (MD) simulation method is applied to fracture and free surface formation in silica glass. Uniaxial strain is applied at different rates to simulated glass structures and the stress is followed as a function of time. The results are found to closely simulate brittle fracture processes by exhibiting a sudden drop in stress to zero at failure. A comparison of results of simulations for different applied strain rates is presented, and it shows major differences in failure stress, strain rate dependence and atomic dynamics prior to and during fracture. Free surfaces formed as a result of the fracture process are analyzed for roughness and compared with experimental measurements. These studies lead to new insights into the processes underlying brittle fracture and surface reconstruction.

Introduction

Brittle fracture is a dynamic, non-linear process that involves a large ensemble of atoms acting cooperatively to relieve the build up of elastic potential energy through atomic motions and the formation of free surfaces. While the initiation of fracture can be predicted by the application of energy balance concepts first developed by Griffith [1], neither the path of the crack, nor its propagation characteristics, nor the structure of the freshly formed surfaces are currently within range of either theory or experiment. The major reason for this lack of understanding is the dominance of local atomic dynamics on crack formation and propagation in the strained solid. Furthermore, the problem is aggravated by the lack of experimental method for examining these atomic dynamics in detail. This type of problem, therefore, can best be investigated by simulation methods like

Molecular Dynamics (MD) whose results can then be compared to the behavior of real materials. MD methods can simulate the atomic trajectories resulting from interatomic interactions. Therefore, they can describe the atomic motions of an ensemble of particles bonded together to form a solid, and how these particles will react to an imposed stress or strain. By examining the details of this response, it is possible to determine some of the processes which govern brittle fracture in solids [2] and the formation of free surfaces. Using an analysis of these surfaces, the results of MD simulations can be compared to experimental data obtained from surface analysis techniques such as Atomic Force Microscopy (AFM) and X-ray Photoelectron Spectroscopy (XPS). Since MD approaches do not simulate chemisorption processes at the crack tip, the studies presented here are limited to processes associated with rapid fracture.

Properties of the rapid brittle fracture phenomenon are obtained from fractography studies which yield structural descriptions of the fracture surface [3], and from fractoemission studies which give descriptions of particle, electron, acoustic and light emission during fracture [4]. Fractography shows regions of mirror, mist and hackle, corresponding to an apparent increasing degrees of roughness on the fracture surface, and a fractal surface structure in all three regions over a scaling range from macroscopic to submicron dimensions [3]. Despite the apparent differences in roughness, however, these 3 regions exhibit the same fractal dimension factor and vary only in the *scale* of the roughness. Maximum crack velocities have been calculated and measured to be between 40 and 60% the speed of the Rayleigh wave in solids.

In this paper, we will examine the atomic dynamics involved in bond failure using MD studies, and determine how they affect the process of free surface formation during rapid brittle fracture in solids. This analysis is limited to flaw-free and non-chemically mediated fracture.

Molecular Dynamics Simulations of Fracture

MD simulations were conducted on an SiO_2 amorphous structure with spatial extent in 3 dimensions and at room temperature (300K). The simulations were conducted using the Verlet algorithm with full periodic boundary conditions. The calculation time step was 0.5 fs as determined previously [5]. The fundamental phonon vibration period was approximately 40 fs. Except when otherwise noted, all calculations were conducted under an isothermal constraint by which all particle velocities were scaled to maintain constant temperature. The glass phase was formed by quenching at 0.25K/ts from a liquid at 12,000K, then annealing for

2,000 ps at 300K. The liquid was formed by melting a cristobalite crystal or by assigning random positions and velocities to the particles of the ensemble.

One of the major drawbacks of MD simulations is the lack of quantum mechanical forces [6-8] and the use of essentially classical interatomic interactions. However, a number of excellent empirical force functions have been used and have yielded structural simulations that reasonably resemble real solids, reproducing accurate crystal symmetries and the radial distribution function of glasses. This is especially true of silica simulations. During this study of silica, we have used several interatomic force functions and compared the results. The comparison between the Soules two-body potential [9] and the Feuston-Garofalini 3-body potential [10] is discussed in Ref. 11. We find that there are differences in detail between the simulations using the different force functions. However, the general behavior and the ensemble averages exhibit little difference. Therefore, we describe below the results of simulations that are reasonably independent of the form of the force function. When differences appear, they will be noted.

One of the great advantages of MD simulations is that the ensemble is simulated in 3 dimensions at a realistic temperature, therefore, the effect of thermal vibrations of the constituent atoms can be studied. The room temperature density for the glass (2.20 g/cm^3) was obtained at atmospheric pressure. The structures produced for silica glass show excellent agreement of the simulated radial distribution function with that derived from neutron diffraction data [9,11].

An uniaxial strain was applied to a box of 2,000 (40 x 20 x 20 angstroms) and 4,000 particles along the long direction, by appropriately scaling the particle positions along that axis. Therefore, the strain was applied uniformly to all particles. Strain rates were used which varied from the speed of sound to about 1/1000 the sound velocity.

Stress-strain behavior was analyzed as a function of strain rate. In all cases, the longitudinal stress developed by the application of the uniaxial strain exhibited an elastic increase up to a maximum, then a decrease to zero. The cohesive strength of the material was assigned to the stress maximum, and failure was assumed to occur as soon as the stress decreased. Sudden collapse of the material and of the induced stress was avoided by the application of a constant strain. In real materials, the fracture process grows catastrophically as the strain runs away at the point of failure under increasing stress at the crack tip.

The application of strain at rates near or above the speed of sound in the material is essentially unreal, however it allows a study of the fracture process in the absence of thermal atomic motions and allows comparison to the static models by keeping the atoms frozen. This is similar to conducting calculations at low temperatures, except that the density does not have to be readjusted. Furthermore, by comparing the high strain rate results with the low strain rate simulations, it is possible to examine the effect of thermal motions on the fracture process.

Figure 1 shows the stress-strain relationships for different strain rates, for simulations using the 2-body, Soules Potential. The simulations show that the elastic region is independent of strain rate below 1.0/ps. There appears to be a hardening of the modulus at larger strain rates. Fracture in the glass at low strain rates appears to be preceded by a region of local plastic deformation where voids are forming and coalescing. In all cases, the stress after failure drops rapidly or eventually to zero indicating a loss of interaction between the two parts of the solid. The failure stress of the glass at low strain rates is about 27 GPa, while that at higher strain rates is 58 GPa. This difference of over 50% is interesting and infers a novel process in the mechanism of fracture as described below.

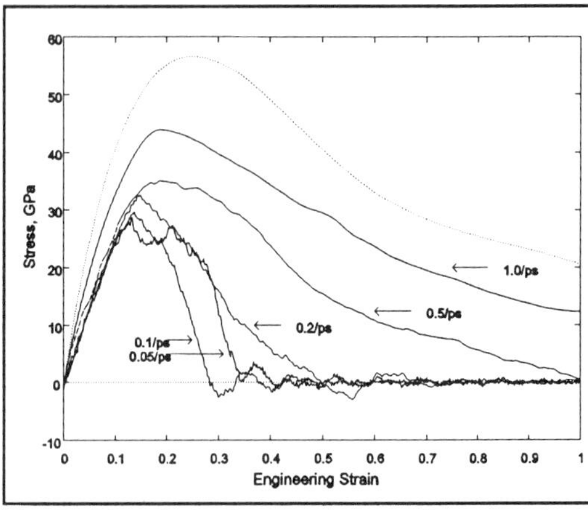

Fig. 1 - Glass behavior at various strain rates using the Soules potential

Interpretation of Collective Atomic Dynamics During Fracture

An examination of the change in bond length and bond angle distribution presents a novel picture of fracture in these systems. For example, at high strain rates, the majority of the applied strain is taken up by bond length changes that are relatively evenly distributed over all the bonds of the material until failure begins to occur. This is particularly pronounced in the 3-body simulation. By contrast, in the low strain rate regime, the applied strain is primarily taken up by bond angle increases rather than any changes in bond length. This change in bond angles involves very small displacements of the surrounding atoms and does not constitute an atomic diffusive process. The process is simply a result of the thermal motions of the atoms at 300K. Yet it appears to be the underlying mechanism for void growth and coalescence.

The glass at high strain rates primarily exhibits bond length increases, indicating that the structure is rigid, however, it fails at a significantly lower stress than a crystal phase of the same composition simulated using the same potentials [11]. This premature failure is due to the presence in the glass of local density fluctuations, over a wide range of sizes, resulting from its high temperature formation and the amorphous nature of its structure. Regions of low density or larger free volume have a lower cohesive energy than the higher density regions, therefore they exhibit a weaker resistance to failure. Failure is thus more localized in the glass and seeks the regions of lower local density. This effect is not present in the crystal, therefore, the crystal only fails when a dynamic instability arises or when the entire structure fails. By contrast, the glass has pre-existing internal flaws which can nucleate failure. This mechanism explains the difference in strengths between the high and low strain rate studies in the glass. For example, at low strain rates, the strain to failure is applied over many thermal fluctuations (acoustic vibration periods). The thermal fluctuations allow the free volume created by the induced strain to flow through the lattice to the regions with the largest voids. This process acts to concentrate the void volume at a single location which eventually fractures the box at a single plane. By contrast, during the rapid application of strain, void diffusion through the lattice is not as facilitated, and large voids grow simultaneously at several locations. Failure of the box occurs over several planes.

These results suggest a mechanism of fracture by which the application of strain to a material is propagated through the matrix according to the compliance of the local structure. When the strain is applied uniformly and slowly compared to the sound velocity, thermal motions rearrange the structure to make it more compliant, thus, as flaws begin to form (eg. structural defects, point or extended defects or grain boundaries), localized weaknesses develop and the strain migrates to these defects. These results postulate a picture in which the strain in real materials, migrates to the strained region ahead of the crack tip and causes voids which eventually coalesce with the propagating crack by means of thermal motion of the atoms, thus forming a smooth crack. At higher strain rates, approaching the Rayleigh velocity, the thermal atomic motions are ineffective in diffusing free volume, and the applied strain causes local voids to grow without coalescence. Branching cracks develop ahead of the crack tip and compete with each other for stress relief.. The fracture surface is rough on a larger scale.

Examination of Fracture Surface

An analysis of the fracture surface is possible by recording the atoms detected by a probe of predetermined size. We used 3 probe sizes: 1, 2.0 and 2.5 Å. The smallest probe penetrates between adjacent oxygen atoms, while the two larger probes do not. Figure 2 shows the results of a surface analysis using the 2.5 Å probe. Table 1 shows the surface roughness measured as a standard deviation from an average surface plane normal to the direction of strain application. Figure 2 shows clearly the predominance of oxygen atoms on the fracture surface, in support of our previous studies on bond fracture which clearly show that silicon atoms penetrate the surface toward the bulk during surface reconstruction [12]. The table shows surface roughnesses of 3-6Å with no noticeable dependence on strain rate. Better agreement is seen between the values obtained with the larger more realistic probes. These results are in good agreement with AFM measurements conducted by Innis et al. [13] on fracture surfaces of silicon dioxide glass with a 300Å probe. The lack of roughness dependence on strain rate is consistent with the fractal nature of these surfaces from the mirror to the hackle region. The difference in visual appearance of these regions is a result of the *scale* of the roughness over larger dimensions than can be accessed by MD studies. Thus on a microscopic scale, the AFM measurements show the same roughness, while if the scale is increased to the submicron (optically accessible) range, then differences in roughnesses appear. We speculate that the roughness exhibited by the MD simulations is the source of the fractal characteristic of the surfaces, and the larger scale differences observed in the mist and hackle regions arise from the multiple fracture surfaces developed at high strain rates in the MD simulation.

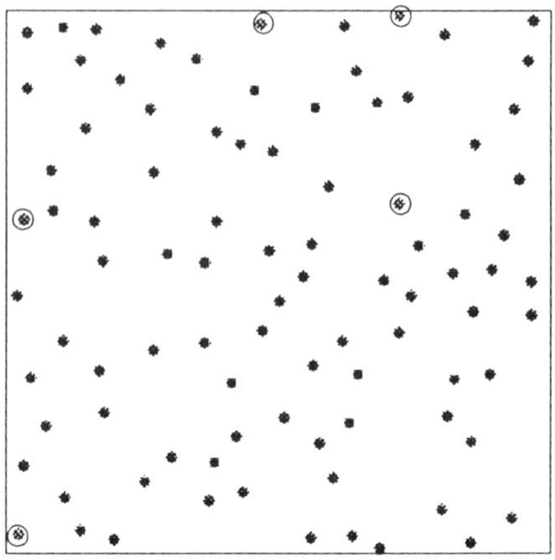

Figure 2 - Fracture surface using a 2.5Å probe showing a predominance of oxygen atoms (solid circles) over silicon atoms (dotted open circles).

Conclusions

MD simulations of the fracture of silicon dioxide glass have provided a basis for the fundamental understanding of the processes underlying rapid brittle fracture and surface reconstruction during fracture. A decrease in structural strength with decreasing strain rate is observed and is related to failure of the solid by coalescence of voids through thermal motions of the atoms. The fracture surface roughness measured agrees well with experimental studies and appears to be independent of strain rate. However multiple fracture planes are developed only at high strain rates, simulating a crack propagating at speeds near the Rayleigh velocity. This is interpreted to indicate that while the surface roughness on the atomic level remains unchanged by crack velocity, the roughness on a larger scale will show dependence on crack velocity, varying from smooth appearance at low velocities to very rough near the Rayleigh velocity. Constancy of surface roughness has been observed in AFM experimental studies of mirror, mist and hackle regions, and differences of roughness on a larger scale, due to crack branching, have been observed to distinguish between mirror, mist and hackle.

Table 1 - MD Surface Roughness Measurements**

Loading Strain Rate	1.0Å Probe	2.0Å Probe	2.5Å Probe
0.01/ps	5.9 ± 0.3 Å	4.0 ± 0.4 Å	3.8 ± 0.5 Å
0.05/ps	5.3 ± 0.5 Å	3.8 ± 0.5 Å	3.6 ± 0.5 Å
0.10/ps	4.1 ± 2.5 Å	3.8 ± 1.4 Å	3.8 ± 1.3 Å
0.20/ps	5.8 ± 0.4 Å	4.5 ± 0.4 Å	4.1 ± 0.4 Å
0.50/ps	6.3 ± 1.0 Å	3.7 ± 0.6 Å	4.6 ± 1.8 Å
1.00/ps	4.2 ± 0.3 Å	4.0 ± 0.2 Å	3.4 ± 1.1 Å

**AFM Measurements by Darryl Innis (AT&T Bell Labs) [13] show that the roughness of silica glass is 3.4Å for a fracture surface and 1.8Å for a melt surface.

References

1. Griffith, AA, Philos Trans R Soc **A221** (1921), 163.
2. Simmons, JH, in Experimental Techniques of Glass Science, Simmons, CJ and El-Bayoumi OH, eds. Am Ceram Soc (1993), 315.

3. Mecholsky JJ, Passoja DE and F.-Ringel KS, J Am Ceram Soc **72** (1989), 60.
4. Langford SC, Ma Z and Dickinson JT, J Mater Res **4** (1989), 1272.
5. Ochoa R, Swiler TP and Simmons JH, J Non-Cryst Solids **134** (1991), 57.
6. Michalske TA and Frieman SW, J Am Ceram Soc **66** (1984), 407.
7. Ng W, White GS and Freiman SW, J Am Ceram Soc **75** (1992), 3097.
8. West JK and Hench LL, J Am Ceram Soc **78** (1995), 1093.
9. Soules TJ, J Non-Cryst Solids **123** (1990), 48.
10. Feuston BP and Garofalini, J Chem Phys **91** (1989), 564.
11. Swiler TP, Simmons JH and Wright AC, J Non-Cryst Solids **182** (1995), 68.
12. Simmons JH, Swiler TP and Ochoa R, J Non-Cryst Solids **134** (1991), 179.
13. Innis D. and Kurkjian CR, Fall Meeting of the Am Ceram Soc, Columbus, OH, Nov. 1994.

Modeling Tetrahedral Edge Shared Defect Sites on Dehydroxylated Silica: A Molecular Orbital Study

Rajiv D. Bendale and Larry L. Hench
University of Florida, Gainesville, FL 32611

ABSTRACT

The characteristic infrared (IR) bands at 908, 888 and 940 cm^{-1} associated with tetrahedral edge shared defect sites on dehydroxylated silica have been investigated using cyclo-disilicic acid and cyclosiloxane molecules as models. The PM3 molecular orbital method applied to these molecules yields IR spectra in good agreement with experiment. Theoretical IR spectra generated from ^{18}O labeled Si-O-Si bridging bonds show isotopic shifts consistent with experiment.

INTRODUCTION

IR absorption spectroscopy has been successfully applied to the analysis of high surface area materials and related adsorption complexes [1-4]. The surface of silica shows a large number of sites available for adsorption of gaseous species, where silanol groups on the surface form the major sites for physisorption. On dehydroxylation of silica in vacuum above 500° C, highly reactive surface sites are created. Extensive infrared (IR) studies of these surfaces indicate that these sites are characterized by two strong IR bands at 908 and 888 cm^{-1} [3-4] and a shoulder at 940 cm^{-1}. These sites can be eliminated by dissociative chemisorption reactions with water and ammonia [3-4]. The structure of these sites or surface defects has been postulated by Michalske and Bunker [5] to consist of strained edge shared silicate tetrahedra generated through the dehydroxylation of adjacent silanol groups. Here we examine two chemical structures containing the strained siloxane bridge as models for these defect sites using the PM3 [6] model Hamiltonian.

METHOD AND MODELS

The PM3 model [6] is based on the Neglect of Diatomic Differential Overlap (NDDO) approximation, and is parameterized to reproduce the geometry and heat of formation of molecules. The equilibrium structures obtained have been subjected to normal coordinate analysis to obtain the IR spectra and are compared with experiment. Earlier studies [7] examined active sites using cyclosiloxane as a model for MO calculation at the ab-initio Hartree-Fock level. In this model each silicon atom has two of its valencies satisfied by hydrogen atoms. Here we consider two molecules (see Figure 1): cyclo-disilicic acid and cyclosiloxane as models for the defect site. The acid structure has silicon atoms tetrahedrally coordinated by oxygen, similar to the environment of a silicon atom in the bulk or on the surface of silica as opposed to the cyclosiloxane [7] where the silicon is coordinated by two oxygen and two hydrogen atoms.

RESULTS AND DISCUSSION

Cyclodisilicic acid has four possible isomers, of which two are of C_{2v} symmetry and the remaining are of C_{2h} and D_{2h} symmetry. Geometry optimization of these isomers using the PM3 model yields the C_{2h} structure (fig.1a) as the most stable isomer (ΔH_f -510 Kcal/mol). The optimized geometry of the two molecules considered is shown in the figures 1a and 1b. Each of the structures reflects

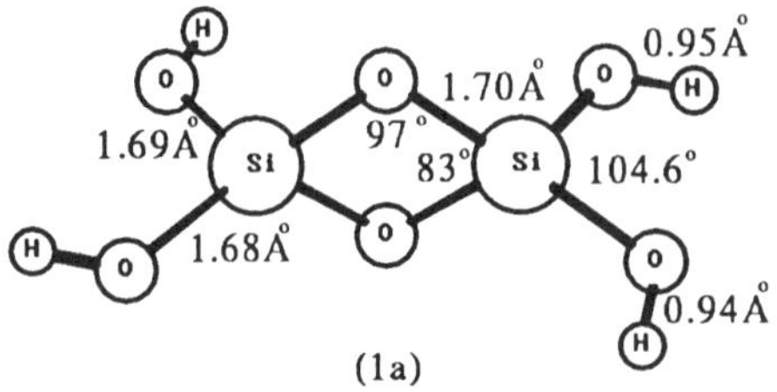

Figure 1: The equilibrium structure of the model compounds (1a) cyclodisilicicacid and (1b) cyclosiloxane optimized using the PM3 model.

the strained ring through the unusually acute O-Si-O bond angle of ~83°, normally ~109.5° in the tetrahedral configuration, and small Si-O-Si bond angle of ~97°, which is ~144° on an average in silica polymorphs. The X-ray crystal structure of two substituted cyclodisiloxanes shows that these acute bond angles do occur naturally. The crystal structure of di-*t*-butyl-dimesityl cyclodisiloxane [7] in D_{2h} symmetry, shows bond angles of Si-O-Si=91.2°, O-Si-O=88.9° and a Si-O bond length of 1.68Å which compares very favorably with the geometry of our models. In contrast the crystal structure of tetra mesityl cyclodisiloxane [8] with C_2 symmetry, shows an opposite trend in the bond angles, where Si-O-Si=86°, O-Si-O=94° and two distinct Si-O bondlengths of 1.66Å and 1.72Å. This structure also shows a twist of 11° about the Si-Si axis, a feature not present in any of the other structures discussed. A comparison between cyclosiloxane (fig. 1b) with D_{2h} symmetry and cyclodisilicic acid (fig. 1a) with C_{2h} symmetry, shows that the Si-O bridging bond length decreases slightly when silicon atom is completely coordinated by oxygen atoms. The O-Si-O bond angle also reflects this decrease. The Si-O-Si bond angle increases. The effect of different chemical environment of silicon atoms at the defect site should be reflected in the IR spectra of the chemical models.

Earlier studies on vacuum dehydroxylated silica indicate the presence of two well defined IR bands at 888 and 908 cm^{-1} associated with the creation of surface defect sites [3,4] These bands disappear when the defect sites react with water vapor, correlated with a corresponding increase in intensity of IR bands due to surface silanol groups. Michalske and Bunker [5] postulated that these defect bands at 888, 908 and 940 cm^{-1} arise from the ring vibrations of strained edge shared silica tetrahedra. We now examine the model structures in this light to determine the origin of these vibrations. The computed IR spectra of cyclodisilicic acid and cyclosiloxane are presented in Table I. This table also shows the isotopic shifts generated by isotopically substituting the bridging oxygen atom. The principal axis (z-axis) for cyclosiloxane is perpendicular to the plane of the ring, with the x-axis passing through the Si atoms. The principal axis for the cyclodisilicic acid molecule is chosen to lie along the bridging O atoms, with the x axis passing through the Si atoms. For both C_{2h} and D_{2h} point groups, only symmetric vibrations with the "g" label are raman active. For the cyclodisilicic acid, the A_u vibrations occur in the plane and

B_u vibrations occur out of the plane of the central $(SiO)_2$ ring. We will primarily focus our attention on the ring vibrations.

Consider the PM3 generated IR spectrum of cyclodisilicic acid which has C_{2h} symmetry. The unscaled frequencies are listed for intensities larger than 0.01 Debye²/amu-Å². The two vibrations at 92, 209 and 272 cm^{-1} with A_u character are largely due to displacements of Si-OH parallel to the plane of the ring with a small component due to in plane oxygen motion. Although the intensity of the vibration at 272 cm^{-1} is large, the contribution from the ring motion is very small. The B_u mode at 347 cm^{-1} is due to strong contribution from out of plane bending of the ring with lesser contribution from hydroxyl groups. The vibrations perpendicular to the ring plane at 782 and 970 cm^{-1} involve in-plane ring motion of the oxygen atoms and Si-OH stretching modes. The 852 (B_g) and 867 (A_u) cm^{-1} modes are associated with pure in-plane ring distortions. The band at 1074 cm^{-1} is also due to in plane displacement of all ring atoms. The higher energy vibrations at 905 and 999 cm^{-1} involve out of plane motion of silicon atoms with some contribution from the Si-OH bending modes. since each of the bands at 347, 782, 852, 867, 970 and 1074 cm^{-1} involve motion of the bridging oxygen atoms, a shift in frequency is observed on isotopic substitution of one of the bridging oxygens, as can be seen from Table 1. In particular the bands at 852 and 867 cm^{-1} compare favorably with a shift of 7 cm^{-1} from experiment for the 888 cm^{-1} defect band. The 9 cm^{-1} shift in the 908 cm^{-1} defect band can probably be compared to the calculated 4 cm^{-1} shift in the band at 970 cm^{-1}. This band as described above is due to ring oxygen motion. In addition Morrow and Cody [3] refer to the presence of a weak shoulder at 940 cm^{-1} in the spectrum of dehydroxylated silica, accompanying the defect bands. This band may be

Table I. The computed IR spectrum of cyclodisilicic acid and cyclosiloxane using the PM3 method. The scaled ab-initio spectrum for cyclosiloxane from [7] is shown for comparison. The computed IR spectrum on isotopic substitution of a bridging oxygen is also presented. The computed intensities are in Debye²/amu-Å².

cyclodisilicic acid			cyclosiloxane			ab-initio [7]	
sym.	cm^{-1}	^{18}O shift cm^{-1}	sym.	cm^{-1}	^{18}O shift cm^{-1}	sym.	cm^{-1}
A_u	99 (0.38)	0	B_{1u}	92 (0.54)	1	B_{1u}	216
A_u	209 (0.19)	1					
B_u	234 (0.62)	1					
A_u	272 (11.66)	0	B_{1u}	607 (1.85)	2	B_{2u}	703
B_u	347 (2.11)	6					
B_u	782 (0.09)	10	B_{2u}	711 (0.23)	5	B_{1u}	719
B_g	852 (0.03)	16	B_{3u}	893 (0.15)	17	B_{3u}	801
A_u	867 (7.63)	13	B_{1g}	898 (0.57)	6	B_{1g}	844
B_u	905 (7.90)	0	B_{2u}	901 (8.00)	8	B_{2u}	961
B_u	970 (6.39)	4					
A_g	971 (0.20)	1	B_{3u}	962 (2.63)	12	B_{3u}	966
B_u	999 (4.06)	0	B_{2g}	1930 (2.69)	0	B_{2g}	2146
B_u	1074 (4.98)	4	B_{1u}	1930 (3.04)	0	B_{3u}	2150
B_u	3897 (0.68)	0	B_{3u}	1950 (3.59)	0	B_{1u}	2151
B_u	3916 (0.84)	0	A_g	1954 (0.27)	0	A_g	2161

ascribed to the computed vibration at 1074 cm^{-1} which also shows an isotopic shift due to the in-plane oxygen displacement.

The PM3 computed IR spectrum of cyclosiloxane shown in Table I is also compared with the ab-initio IR spectrum by Bunker et al. [7]. Only vibrations with non zero intensities have been listed. The bands at 92 and 607 cm^{-1} arise from ring distortions perpendicular to the ring plane. The B_{2u} modes at 711 and 901 cm^{-1} and B_{3u} mode at 962 cm^{-1} involve motion of ring atoms in the ring plane. The 893 cm^{-1} and the band at 1930-1954 cm^{-1} arise due to Si-H vibrations. The PM3 computed frequencies tend to underestimate the vibrational energy in comparison with the scaled ab-initio frequencies of Bunker et al. [3]. The computed isotopic shift of 8 cm^{-1} for the 901 cm^{-1} band corresponds best to the experimentally observed 9 cm^{-1} shift for the 888 cm^{-1} defect band. The 908 cm^{-1} defect band would best be assigned to the higher energy B_{3u} mode at 962 cm^{-1}. The isotopic shift then compares very favorably with the experimentally observed one at 8 cm^{-1} . It is difficult to ascribe the weak shoulder observed at 940 cm^{-1} to any vibration for the cyclosiloxane model, which was seen in the cyclodisilicic acid model. It is interesting to see if one can obtain trends in the relative reactivity of gaseous reactants like H_2O and CH_3OH with the cyclo disilicic acid model.

We have identified the transition state in the gas phase chemisorption reaction between cyclo disilicic acid and (a) H_2O and (b) CH_3OH, using the same PM3 model. The structure of these states is depicted

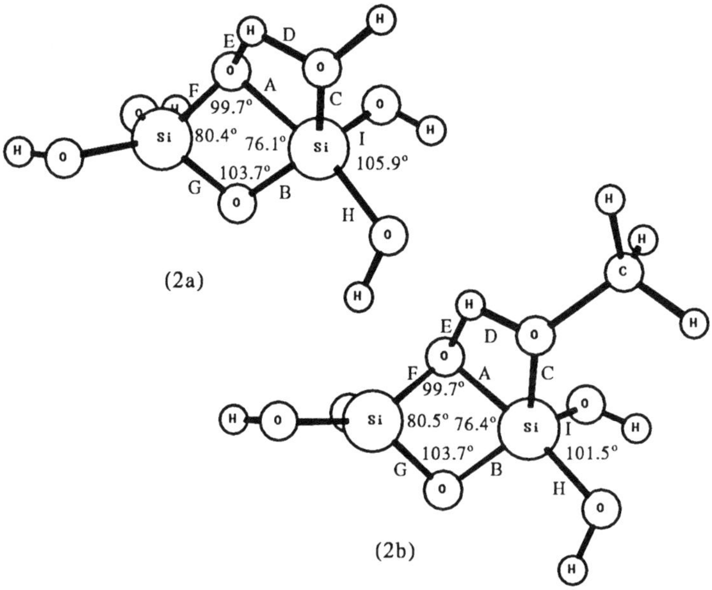

Figure 2: The transition state in the gas phase reaction between cyclodisilicic acid and (2a) H_2O (TS1) and (2b) CH_3OH (TS2) obtained using the PM3 model. Some bond angles are shown for comparison, the alphabetic labels refer to bonds described in Table III.

in Figure 2 along with some of the salient bond lengths and bond angles. The transition state TS1 between cyclodisilicic acid and water has a single imaginary frequency of 1906i cm^{-1} and the transition state TS2 between the acid and methanol has a single imaginary frequency of 2042i cm^{-1}, indicating that the structures are true first order transition states. Table II presents the enthalpy for the path followed from the reactants to the transition state to the ring opened products.

Table II. Computed enthalpies for the gas phase chemisorption reaction between cyclodisilicic acid and (a) water and (b) methanol and the resultant ring opened products computed using the PM3 method. The enthalpy is reported in Kcal/mol.

	ΔH(gas)
$H_2O + (OH)_4(SiO)_2 \rightarrow TS1$	+15
$TS1 \rightarrow (OH)_3SiOSi(OH)_3$	-42
$CH_3OH + (OH)_4(SiO)_2 \rightarrow TS2$	+19
$TS2 \rightarrow (OH)_3SiOSi(OH)_2OCH_3$	-44

The reported values are not corrected for zero point vibrational energy. Michalse and Bunker [5] report kinetic studies which show that methanol is slower to dissociate on the strained siloxane bridging bond of the defect site, than water. This is agreement with our results. Table II shows that the barrier to reaction of cyclodisilicic acid with water is 4 Kcal/mol lower than for methanol. An explanation for lower reaction barrier for Si-O bond opening by water as compared to methanol may be seen from Table III. Here, the bond order for the O-H bond (E) between H of water and the bridging O of the acid

Table III. Bond lengths and bond orders of some of the bonding atoms in the transition state for the reaction of cyclodisilicic acid (R1) and (a) water (TS1) and (b) methanol (TS2). The bond lengths in Å, are shown in parenthesis. See Figures 2a and 2b for bonds referred to in table headings.

	A	B	C	D	E	F	G	H	I
TS1	0.56	0.76	0.52	0.46	0.40	0.78	0.93	0.88	0.83
TS2	0.57	0.76	0.49	0.47	0.39	0.78	0.93	0.89	0.83
defect	0.85	0.85				0.85	0.85	0.93	0.8

is slightly lager than that of methanol. The corresponding bond length of the breaking O-H bond (D), as seen from figure 2 and Table III, for water is more extended than in methanol. This implies that the transition state for water has a weakened OH bond in H_2O, which should dissociate and open the ring more readily than in the case for methanol. The O-Si-O bond angles in the transition state TS1 also indicate a more strained ring system than the TS2 structure which also points toward a faster dissociation of TS1 over TS2.

CONCLUSIONS

The cyclodisilicic acid and cyclosiloxane model chosen to model the edge shared tetrahedral defect sites on dehydroxylated silica support the postulated structure of the defect site. The cyclodisilicic acid IR spectrum matches the experimentally observed defect bands at 888 and 908 cm^{-1} and the associated weaker shoulder at 940 cm^{-1}. The computed ^{18}O labelled shifts also agree favorably with the experimental ones, however no experimental information is available for the 940 cm^{-1} shoulder. We predict a 4 cm^{-1} shift for this band. The transition state calculations show that gas phase water and methanol have similar reaction barriers with a 4 Kcal/mol lower barrier predicted for water as compared to methanol, borne out by the kinetic studies reported in literature.

REFERENCES

[1] L.H. Little, "Infrared Spectra of Adsorbed Species" Academic Press, New York, 1966

[2] P. Hoffmann and E. Knozinger, "Novel aspects of mid and far IR fourier spectroscopy applied to surface and adsorption studies on SiO_2, Surf. Sci. **188**, 181-198 (1987).

[3] B.A. Morrow and I.A. Cody, "Infrared studies of reactions on oxide surfaces. 5. lewis acid sites on dehydroxylated silica", J. Phys. Chem. **80**, 1995-1998 (1976)

[4] B.A. Morrow and I.A. Cody, "Infrared studies of reactions on oxide surfaces. 6. Active sites on dehydroxylated silica for chemisorption of ammonia and water", J. Phys. Chem. **80**, 1998-2004 (1976)

[5] T.A. Michalske and B.C. Bunker, "Slow fracture model based on strained silicate structures", J. Appl. Phys. **55**, 2686-2693 (1984).

[6] J.J.P Stewart, J. Comp. Chem. **10**, 209 (1989); J.J.P. Stewart, J. Comp. Chem. **10**, 221 (1989).

[7] B.C. Bunker, D.M. Haaland, K.J Ward, T.A. Michalske, W.L. Smith, J.S. Binkley, C.F. Melius and C.A. Balfe, "Infrared spectra of edge shared silicate tetrahedra", Surf. Sci. **210**, 406-428 (1989).

[8] M.J. Fink, K.J. Haler, R. West and J. Michl, "Tetramesityl cyclodisiloxane: A cyclic siloxane with an unusual structure", J. Am. Chem. Soc. **106**, 822-823 (1984).

RECONSTRUCTION OF SURFACES IN NiO

M. Yan and S. P. Chen
Theoretical Division, Los Alamos National Laboratory, Los Alamos, NM 87545

ABSTRACT

We studied the reconstructions of surfaces in NiO by atomistic simulations which utilize Buckingham short range potentials and the shell model. It was found that {hk0} surfaces prefer to reconstruct into {100} surface facets which has the lowest energy. The interaction between these {100} facets is repulsive and converges to zero as size of facets grows. The {111} surface can be stabilized by reconstruction into {100} micro-facets.

INTRODUCTION

Reconstruction of surface has been found to be a common feature in metallic systems, from both experiments and theoretical calculations in the recent years [1-3]. The study of reconstruction in oxide ceramics, however, is still very limited. In this work we studied the reconstruction of {hk0} and {111} surfaces in NiO, which has the NaCl structure. From experiments it has been observed that the {110} and {111} surfaces in MgO, which also has the NaCl structure, facet into {100} surfaces [4].

The interionic forces in NiO have been described by Coulomb long range and Buckingham short range potentials, and the latter has been combined with the shell model to include polarization of anions [5], i.e. an O ion is treated as a positive core and a negative shell connected harmonically by a spring. The Buckingham potentials have been fitted to properties of NiO by Grimes, et al, which include the lattice parameter, the lattice energy, the elastic constants and the dielectric constants [6, 7].

Surface relaxation was carried out using the program MARVIN [8]. The simulation method is similar to what has been used previously [9, 10].

RESULTS AND DISCUSSIONS

1.Reconstruction of {hk0} surfaces

A {hk0} surface in NaCl lattice is a non-polar surface, and the calculated surface energies, γ, are illustrated in Fig.1 as the function of the rotation angle, θ, from the (010) surface It can be seen that the dependence of γ on θ is monotonic, both before and after relaxation. The variation of γ with θ is quite large, ranging from the minimum value of $1.17 J/m^2$ for the relaxed (010) surface to the maximum value of 2.98J/m^2 for the relaxed (110) surface. This is consistent with the experimental observation that in NiO and other metal oxides with the NaCl structure cleavage always takes place along the {100} plane [11]. The reason for the monotonic dependence between the surface energy and the surface orientation has been explained before [7, 12, 13].

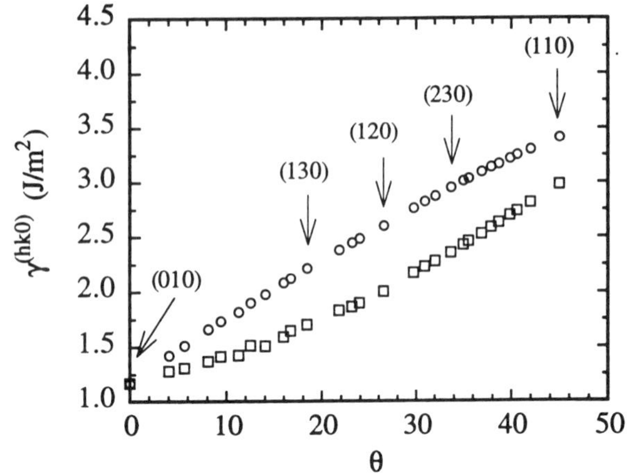

Fig.1. The calculated surface energy, γ, as a function of the rotation angle, θ.

Fig.2(a) shows the structure of the (110) surface projected onto the (001) plane. The black and white circles represent Ni ions and O ions, respectively. Behind each Ni ion along the [001] direction there is an O ion and vice versa.

If the unit cell is chosen to be 2x1, i.e. with dimensions [110]x[001], and ions on alternate [001] columns are removed from the outermost layer, the resulting structure is shown in Fig.2(b). This corresponds to the (110) surface facetting into two {100} surfaces. The relaxed surface energy is 2.00J/m^2, which is much smaller than that without reconstruction, so the reconstruction is favored. Such missing-row reconstruction shown in Fig.2(b) has also been found at surfaces in f.c.c. metals, although the direction of the missing-rows is different [14]. Furthermore, if in general the unit cell is chosen to be mx1, where m is an integer, then following a similar procedure as for the 2x1 unit cell to remove some columns of ions to create {100} surface facets, it is found that the larger the m, the smaller the relaxed surface energy, as is illustrated in Fig.3. In this figure, ρ is the density of {100} facets in the $[1\bar{1}0]$ direction and equals $\sqrt{2}/m$, γ^{mx1} is the relaxed surface energy of the reconstructed (110) surface with unit cell mx1. The scattered symbols are from direct calculations, and the solid curve is from functional fitting, which leads to:

$$\gamma^{mx1}=1.66+0.10\rho+11.17\rho^2 (J/m^2)$$

The first term, 1.66J/m^2, is the ideal energy value of the reconstructed surface if interactions between {100} facets are ignored. The fitting indicates that the interaction between facets is repulsive, and increases with increasing density of facets. Such interaction vanishes gradually as the size of the faceted surface grows, i.e. as ρ->0. The reconstructed surface structure with unit cell 5x1 is shown in Fig.2(c).

Similar reconstructions of {hk0} surfaces into {100} facets have also been found for other surfaces, such as the (120), (130) and (230) surfaces. Their surface energies without and with reconstructions, denoted as γ and γ', respectively, are summarize in Table 1.

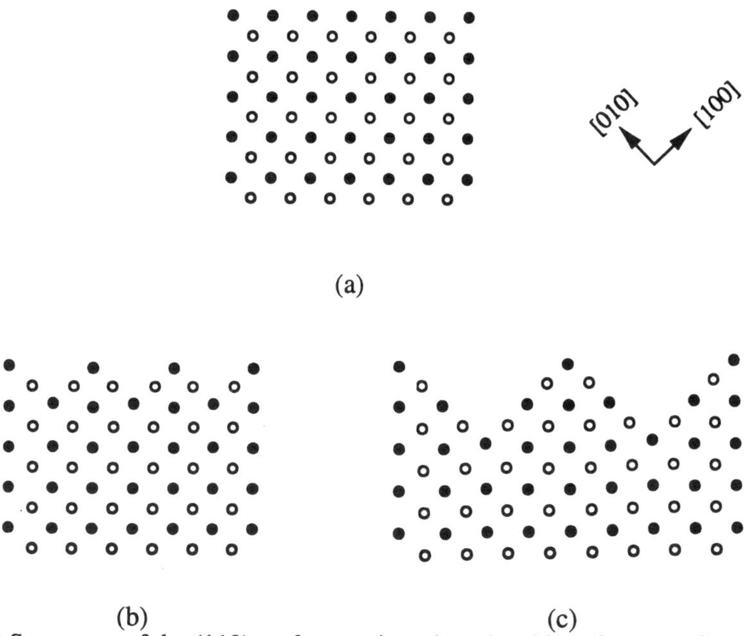

Fig.2.Structures of the (110) surface projected on the (001) plane; (a) flat surface; (b) reconstructed one with unit cell 2x1; (c) reconstructed one with unit cell 5x1.

2. Reconstruction of the (111) surface

The {111} surface is a polar surface in NiO structure, and is not stable in its perfect state [15]. The projected structure of this surface on the (111) plane is shown in Fig.4. The shaded circles represent Ni ions and the white circles O ions. One way to stabilize a (111) surface is to create it in a way such that the structure terminates with only complete octopolar $(NiO)_4$ molecules [16, 17]. Each $(NiO)_4$ molecule is the basis of the NiO lattice if its structure is treated as simple cubic, as shown in Fig.5. The (111) surface created in this way is shown in Fig.6(a) as

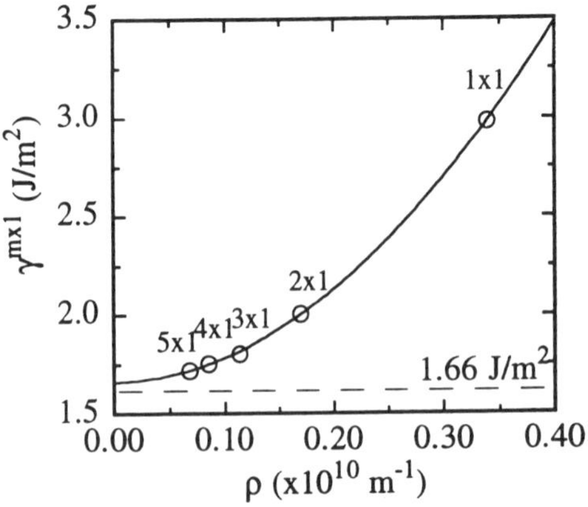

Fig.3 Energies of reconstructed (110) surfaces as a function of density of {100} facets in the [110] direction.

	$\gamma(J/m^2)$	$\gamma'(J/m^2)$
(120)	2.002	1.680
(130)	1.700	1.542
(230)	2.352	1.833

Table 1. Energies for some other {hk0} surfaces without and with reconstructions

projected on the (111) plane, and in Fig.6(b) as viewed from an oblique angle. Such reconstructed (111) surface has no dipole moment along the surface normal, because the dipole for a complete octopolar $(NiO)_4$ molecule is zero. The (111) surface structure can thus be stabilized. Comparing Figs.6(a)(b) with Fig.4, it can be seen that the perfect flat {111} surface reconstructs into a ragged structure which can also be considered as micro-facets of the {100} surface. The surface energy is found to be $5.25J/m^2$ and $2.62J/m^2$ before and after relaxations, respectively. If the interactions between the {100} facets is ignored, the surface energy should be $\sqrt{3}\gamma^{\{100\}}$, which equals $2.03J/m^2$. So the results indicate that, similar to the (110) surface, the interaction between {100} facets on the (111) surface is repulsive, and should be expected to decrease as the {100} micro-facets grow. Such a result is consistent with the LEED observation that, in MgO, which has the same structure as NiO, (111) surface facets into {100} surfaces [4]. It also suggests an alternative explanation other than impurity segregations for the observation that NiO can be stabilized into {111} surface [18].

CONCLUSIONS

Energy of a {hk0} surface increases in the monotonic way with the increasing rotation angle from the (010) surface. The {hk0} surfaces with higher energies prefer to reconstruct into {100} surface facets. The interaction between these {100} facets is repulsive and converges to zero as the size of facets grows. The (111) surface can be stabilized by reconstruction into many {100} micro-facets, and the relaxed surface energy is 2.62 J/m^2.

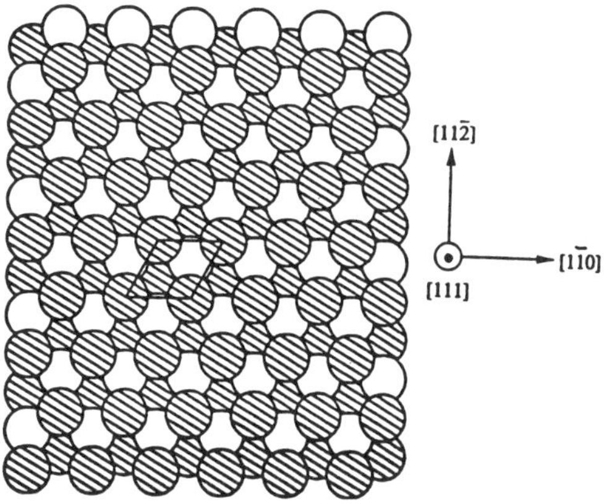

Fig.4. The structure of the perfect (111) surface projected on the (111) plane.

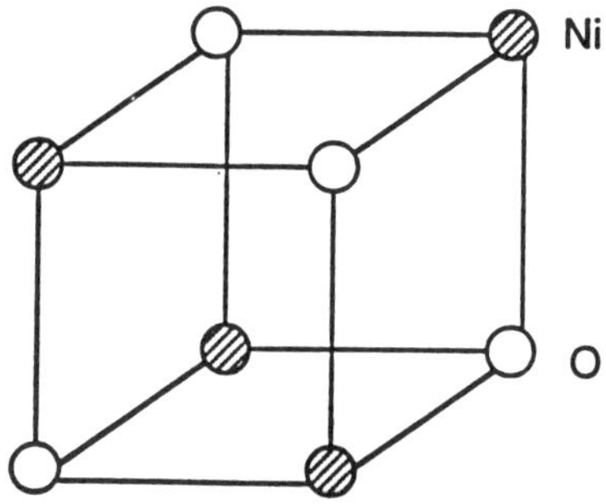

Fig.5. The structure of the octopolar $(NiO)_4$ molecule.

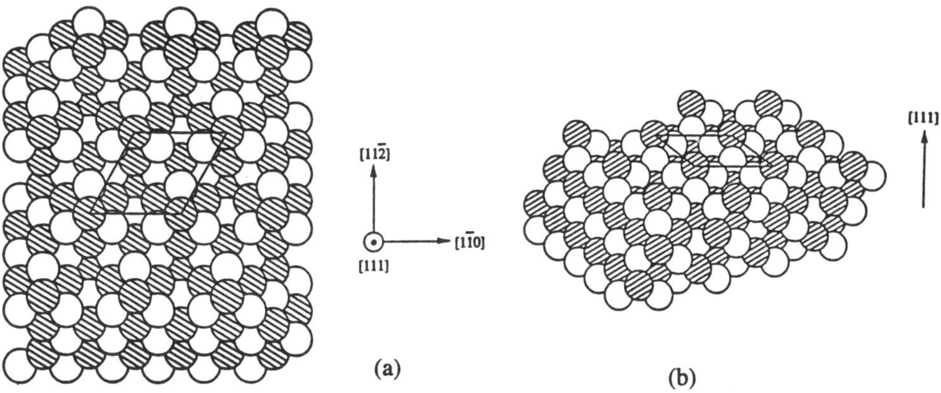

Fig.6. Structures of the reconstructed (111) surface; (a) projected on the (111) plane; and (b) viewed obliquely.

ACKNOWLEDGMENTS

We acknowledge Drs. R. W. Grimes and S. Vyas for providing their potentials, and acknowledge Dr. D. H. Gay for helps regarding MARVIN program. We also thank Dr. A. F. Voter for helpful discussions. This research was supported by Division of Materials Science, Office of Basic Energy Science of Department of Energy.

REFERENCES

1. M. A. Van Hove and S. Y. Tong, eds. *The Structure of Surfaces*, Vol.2 of Springer Series in Surface Sciences (Springer, Berlin, 1985).
2. S. P. Chen, A. F. Voter and D. J. Srolovitz, *Oscillatory Surface Relaxations in Ni, Al and Their Ordered Alloys,* Phys. Rev. Lett. 57, 1308 (1986).
3. M. S. Daw and S. Foiles, *Order-Disorder Transition of Au and Pt (110) Surfaces*, Phys. Rev. Lett. 59, 2756 (1987).
4. V. Henrich, *Thermal Facetting of (110) and (111) Surfaces of MgO*, Surf. Sci. 57, 385 (1976).
5. B. G. Dick and A. W. Overhauser, *Theory of the Dielectric Constants of Alkali Halide Crystals*, Phys. Rev. 112, 90 (1958).
6 R. Grimes and S. Vyas, *private communications* (1994).
7. M. Yan, S. P. Chen, T. E. Mitchell, D. H. Gay, S. Vyas and R. W. Grimes, *Atomistic Studies of Energies and Structures of (hk0) Surfaces in NiO*, to be published in Phil. Mag. A. (1995).
8. D. H. Gay and A. L. Rohl, *MARVIN-A New Computer Code for Studying Surfaces and Interface and its Application to Predicting Crystal Morphology, submit to Faraday Transaction* (1994).
9. D. E. Parry, *The Electrostatic Potential in the Surface Region of an Ionic Crystal*, Surf. Sci. 49, 433 (1975).
10 P. W. Tasker, *The Surface Energies, Surface Tension and Surface Structure of the Alkali Halide Crystals,* Philo. Mag. A 39, 119 (1979).
11. V. E. Henrich and P. A. Cox, *The Surface Science of Metal Oxides*, Cambridge University Press, (1994).
12. P. W. Tasker and D. M. Duffy, *The Structure and Properties of the Stepped Surfaces of MgO and NiO*, Surf. Sci. 137, 91 (1984).
13. D. J. Srolovitz and J. P. Hirth, *Elastic Analysis of the Energy and Relaxation of Stepped Surfaces,* Surf. Sci. 255, 111 (1991).
14. S. P. Chen and A. F. Voter, *Reconstruction of the (310), (210) and (110) Surfaces in fcc Metals,* Surf. Sci. Lett. 244, L107 (1991).
15. P. W. Tasker, *The Stability of Ionic Crystal Surfaces*, J. Phys. C: Solid State Phys. 12, 4977 (1979).
16. R. Lacman, *Colloq. Int. C. N. R. S.* 152, 195 (1965).
17. D. Wolf, *Reconstruction of NaCl Surfaces from a Dipole Solution to the Madelung Problem,* Phys. Rev. Lett. 68, 3315 (1992).
18. C. A. Handwerker, M. D. Vaudin and J. E. Blendell, *Equilibrium Crystal Shape and Surface Phase Diagram at Surfaces in Ceramics*, Journal de Phys. C5, 367 (1988).

SURFACE PHASE TRANSITIONS IN CaO-FeO AND CaO-NiO

C. Battaile, R. Najafabadi*, and D.J. Srolovitz

Dept. of Materials Sci. and Eng.; University of Michigan; Ann Arbor, MI 48109
*Knolls Atomic Power Laboratory; Schenectady, NY 12309

ABSTRACT

Equilibrium atomic configurations at (011) free surfaces in CaO-FeO and CaO-NiO are predicted using Free Energy Minimization. Conventional pair potentials are used to describe atomic interactions. Vibrational effects are included via the Local Harmonic approximation, and solid solutions are represented by the mean-field Effective Atom model. Using these techniques, phase transitions are predicted at (011) free surfaces at T = 2000K. This approach can be used to predict the entire surface phase diagram.

INTRODUCTION

The properties of free surfaces are important in a number of applications, e.g., sintering[1] and catalysis[2]. Complete surface characterization on an atomic scale is usually difficult, and computer simulations[3] have proven valuable for predicting surface properties.

Several atomistic simulation techniques are available for surface studies. Monte Carlo methods[4] provide a powerful tool for predicting equilibrium configurations. Molecular Dynamics[5] allows the simulation of both equilibrium states and kinetic processes.

Both of these methods, while accurate, are very computationally demanding. Alternatively, the Free Energy Minimization method[6,7] can be used to efficiently predict equilibrium configurations, and allows direct determination of thermodynamic quantities.

Free Energy Minimization methods have been successfully used to study segregation to surfaces in metals[8,9]. Previous studies of surface segregation in ionic solids[10-14] have used the Langmuir-McLean theory. The Langmuir-McLean theory for the surface concentration as a function of temperature and bulk concentration is usually parameterized on the basis of atomistic simulation determination of the enthalpy of segregation. Although this approach neglects some important temperature effects (i.e., non-configurational entropy), it is useful for predicting solute concentrations on the surface.

In this paper, we use Free Energy Minimization to predict equilibrium atomic configurations and concentration profiles near (011) free surfaces in CaO-FeO and CaO-NiO. This method explicitly includes the effects of temperature, and allows determination of the solute distribution throughout the entire surface region. Strong segregation and evidence for surface phase transitions are observed at T = 2000K which are expected to significantly affect surface properties and behavior.

METHOD

Equilibrium atomic configurations are determined by minimizing the free energy. The foundation of this free energy is the interionic potential. We employ a Born-Mayer[3] form for this. Interionic potential parameters are adopted from Ref. 14. Ewald's Method[15] is used to overcome the difficulties associated with convergence of Coulombic interactions, and the Shell Model[16] is incorporated to account for ionic polarization.

Vibrational effects are approximated using the Local Harmonic model[6]. In this model, atoms are assumed to vibrate as Quasi-Harmonic oscillators, but the vibration of an atom is coupled only to its own vibrations, and not to the vibrations of the other atoms. Within this approximation, the system of $3N$ dynamical equations that describe a collection of N oscillators reduces to N systems of 3 dynamical equations. These can be easily solved to obtain the $3N$ local normal-mode vibrational frequencies of the N oscillators. The vibrational energy (which includes kinetic energy and vibrational entropy) can then be calculated directly from these local frequencies.

Solid solutions are described using the mean-field Effective Atom method[17], which assigns a concentration to each atomic site. This concentration represents a time-average of the site occupancy in a solid solution. Temporal correlations between site concentrations are neglected. The configurational entropy is calculated using a simple point approximation.

We consider a reduced Grand Canonical ensemble in which the total number of ions is fixed, but the relative amounts of solvent and solute species are allowed to vary. The appropriate free energy for this ensemble is the reduced Grand Potential, Ω:

$$\Omega = U + A_\nu - \mathrm{TS_c} + \Delta\mu \sum_{i=1}^{N} c_i^{(A)} \tag{1}$$

where U is the potential energy, A_ν is the vibrational energy[6], T is the temperature, S_c is the configurational entropy[17], $\Delta\mu$ is the difference between the chemical potentials of the solvent and solute, N is the number of atoms in the system, and c_i is the solute concentration at site i. $\Delta\mu$ is set *a priori* and kept constant so that the bulk crystal maintains the desired concentration[17].

RESULTS

The reduced Grand Potential provides the free energy of a solid solution at finite temperature. Equilibrium configurations are determined by minimizing this free energy with respect to atomic positions and site concentrations using the Polak-Ribiere conjugate-gradient algorithm.

To study binary oxide solid solutions, we consider substitution of solute cations onto the solvent cation sublattice. Since CaO-FeO and CaO-NiO are considered in this work, these solid solutions correspond to the substitution of Fe^{2+} or Ni^{2+} onto the Ca^{2+} sublattice in CaO. The Effective Atom site concentrations represent the fraction of cation sites which are occupied by solute cations.

A simulation cell of dimensions 1x1x9 rocksalt unit cells (72 ions) is used. Periodic boundary conditions are imposed in all three directions. Free surfaces are created by increasing the repeat length in the z ([011]) direction by 40 times the nearest-neighbor separation. The lattice parameter is set to the equilibrium value for the bulk crystal. The repeat lengths in the x ([100]) and y ([01$\bar{1}$]) directions are fixed such that the surface area remains constant.

Concentration profiles at the (011) surfaces of $(Fe_{3.25}Ca_{96.75})O$ and $(Ni_{0.9}Ca_{99.1})O$ at T = 2000K are shown in Fig. 1a. Solvent (Ca^{2+}) segregates strongly to the first surface atomic layer, and solute (Fe^{2+} or Ni^{2+}) to the second. At <u>very</u> slightly higher bulk solute concentrations, as shown in Fig. 1b, segregation of solute to the second surface layer drastically increases.

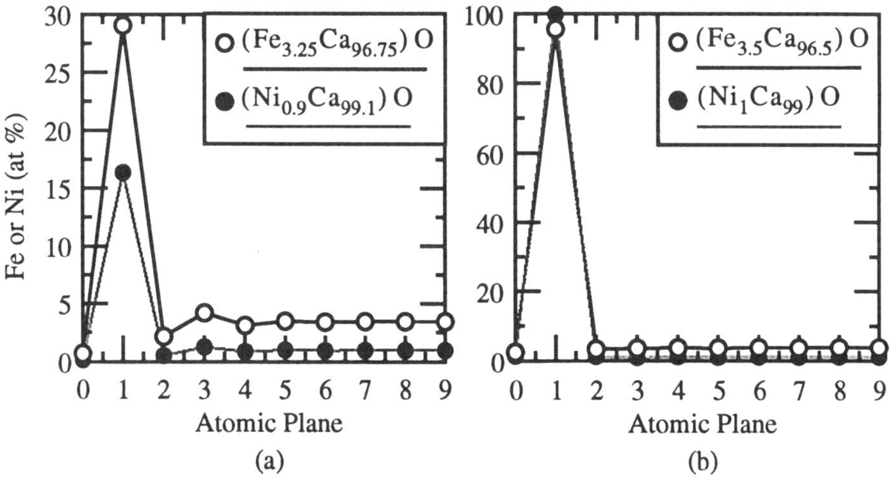

Figure 1. Equilibrium segregation profiles at (011) surfaces in CaO-FeO and CaO-NiO at T = 2000K: (a) $(Fe_{3.25}Ca_{96.75})O$ and $(Ni_{0.9}Ca_{99.1})O$; and (b) $(Fe_{3.5}Ca_{96.5})O$ and $(Ni_1Ca_{99})O$.

DISCUSSION

Segregation to a free surface occurs because the crystal environment at the surface differs from that in the bulk. Reduction of surface energy and relief of misfit strain are primarily responsible for this[18]. The cohesive energies of metal-oxides are relatively large in magnitude, and the differences between the cohesive energies of different oxide systems can therefore be substantial (as manifested in the corresponding surface energy differences). Therefore, the species which segregates to the surface in binary oxides is typically that which has the lower oxide surface energy. This is because of "broken bonds" at the surface.

Furthermore, the solute species is often found to segregate to a surface to minimize strain energy, since the size of a solute atom usually differs significantly from that of the solvent. This effect is primarily responsible for segregation to <u>sub-surface</u> layers in these systems, where there are no significant broken bonds.

The (011) surface free energies of CaO, FeO, and NiO at T = 2000K are 1820, 3150, and 2780 ergs/cm^2, respectively, using the present potentials. The ionic radii[19] of Ca^{2+}, Fe^{2+}, and Ni^{2+} are 1.00, 0.65, and 0.69 Å, respectively. Since CaO has a lower (011) surface free energy than either FeO or NiO, surface energy drives the segregation of Ca^{2+} to the first surface layer in CaO-FeO and CaO-NiO, even though Fe^{2+} and Ni^{2+} are significantly smaller than Ca^{2+} (see Fig. 1).

The first (011) surface layer in dilute CaO-FeO and CaO-NiO relaxes <u>toward</u> the bulk by approximately 10% of the nearest-neighbor separation at T = 2000K. The second layer relaxes <u>away from</u> the bulk by 2%, and the third relaxes <u>toward</u> the bulk by 2%. This large first-layer relaxation, and the occupation of the first layer by the larger Ca^{2+} ion, helps to drive segregation of the smaller solute (Fe or Ni) to the second layer, as seen in Fig. 1.

The sudden increase in the solute concentration on the second surface layer when the bulk concentration is very slighly changed (see Fig. 1) is the result of a surface phase transition. This can be verified by examining the change in surface solute concentration with bulk concentration shown in Fig. 2. The bulk concentration can be mapped directly into the chemical potential, and a discontinuous change in concentration with chemical potential indicates the crossing of a two-phase boundary in a phase diagram. In Fig. 2, the second-layer concentration changes abruptly with bulk concentration (and therefore with chemical potential) at bulk concentrations of about 3.4 at% in CaO-FeO, and 0.9 at% in CaO-NiO at T = 2000K. This indicates that the (011) surfaces undergoes a phase change in dilute CaO-FeO and CaO-NiO at this temperature.

Using this approach, the (011) surface solid solubility limits in CaO-FeO and CaO-NiO at T = 2000K can be determined from Fig. 2. These solubility limits are the surface concentrations at which the discontinuities in Fig. 2 begin and end. The (011) surface solid solubility limits of Fe^{2+} in CaO, Ca^{2+} in FeO, Ni^{2+} in CaO, and Ca^{2+} in NiO at 2000K were determined to be 29.0, 4.4, 16.4, and 0.3 at%, respectively. To extend the analysis, one could obtain these surface solubility limits

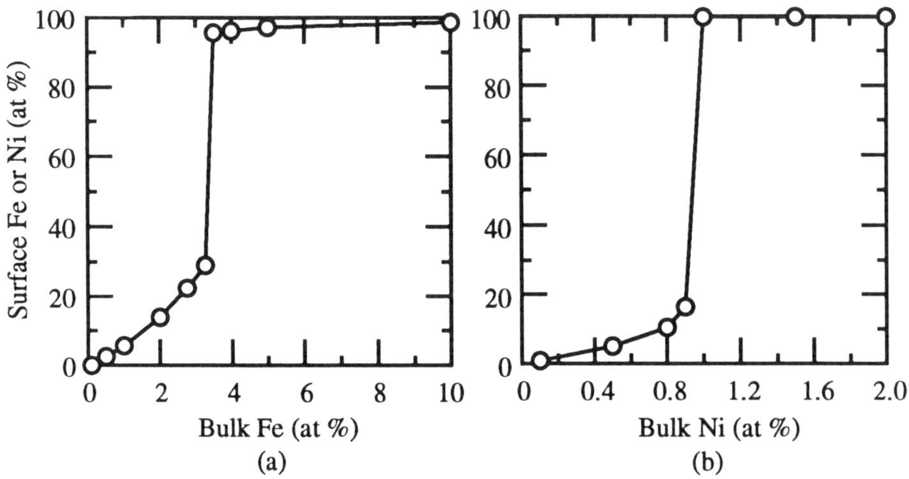

Figure 2. Solute concentration on the second (011) surface layer vs. bulk concentration showing (011) surface phase transitions at T = 2000K in (a) CaO-FeO and (b) CaO-NiO.

at different temperatures to determine the (011) surface phase diagrams of CaO-FeO and CaO-NiO.

Similar transitions have been predicted[9] at the (001) surface of Ni-Pt using these methods. Surface phase changes such as these can significantly affect surface properties[2]. In addition, strong surface segregation like that shown in Figs. 1 and 2 could have considerable impact on surface composition analysis, yielding results which depend strongly on the nature of the analysis technique.

SUMMARY

Equilibrium atomic configurations at (011) surfaces in CaO-FeO and CaO-NiO have been determined using Free Energy Minimization. This method uses conventional pair potentials to describe atomic interactions, the Local Harmonic approximation to include vibrational effects, and the mean-field Effective Atom model to represent solid solutions. Using these techniques, phase transitions have been predicted at (011) free surfaces at T = 2000K. This approach can be used to predict the entire surface phase diagram.

ACKNOWLEDGMENT

This research is supported by the Advanced Research Projects Agency and the Office of Naval Research under Grant No. N00014-91-J-4019.

REFERENCES

1 W.C. Johnson, "Grain Boundary Segregation in Ceramics," *Met. Trans. A*, **8A**, 1413-22 (1977).

2 C. Xu, Y.-L. Tsai, and B.E. Koel, "Adsorption of Cyclohexane and Benzene on Ordered Sn/Pt(111) Surface Alloys," *J. Phys. Chem.*, **98**, 585-93 (1994).

3 *Computer Simulation of Solids*. Edited by C.R.A. Catlow and W.C. Mackrodt. Springer-Verlag, Berlin, 1982.

4 S.M. Foiles, "Calculation of Grain-Boundary Segregation in Ni-Cu Alloys," *Phys. Rev. B*, **40**, 11502-6 (1989).

5 P.J. Mitchell and D. Fincham, "Shell Model Simulations by Adiabatic Dynamics," *J. Phys.: Cond. Matt.*, **5**, 1031-8 (1993).

6 R. LeSar, R. Najafabadi, and D.J. Srolovitz, "Finte-Temperature Defect Properties from Free-Energy Minimization," *Phys. Rev. Lett.*, **63**, 624-7 (1989).

7 A.P. Sutton, "Temperature-Dependent Interatomic Forces," *Phil. Mag. A*, **60**, 147-59 (1989).

8 S.M. Foiles, "Calculation of the Surface Segregation of Ni-Cu Alloys With the Use of the Embedded-Atom Method," *Phys. Rev. B*, **32**, 7685-93 (1985).

9 R. Najafabadi and D.J. Srolovitz, "Order-Disorder Transitions at and Segregation to (001) Ni-Pt Surfaces," *Surf. Sci.*, **286**, 104-15 (1993).

10 P.W. Tasker, E.A. Colbourn, and W.C. Mackrodt, "Segregation of Isovalent Impurity Cations at the Surface of MgO and CaO," *J. Amer. Ceram. Soc.*, **68**, 74-80 (1985).

11 W.C. Mackrodt and P.W. Tasker, "Segregation Isotherms at the Surfaces of Oxides," *J. Amer. Ceram. Soc.*, **72**, 1576-83 (1989).

12 E.A. Colbourn, W.C. Mackrodt, and P.W. Tasker, "The Segregation of Calcium Ions at the Surface of Magnesium Oxide," *J. Mat. Sci.*, **18**, 1917-24 (1983).

13 W.C. Mackrodt, "Atom Lattice Simulations of the Structure, Energetics, and Impurity Segregation Behaviour of Ceramic Boundaries," *Mat. Sci. Res.*, **21**, 271-80 (1987).

14 M.J.L. Sangster and A.M. Stoneham, "Calculations of Off-Centre Displacements of Divalent Substitutional Ions in CaO, SrO, and BaO from Model Potentials," *Phil. Mag. B*, **43**, 597-608 (1981).

15 P.P. Ewald, "Die Berechnung optischer und elektrostatischer Gitterpotentiale (Calculation of Optical and Electrostatic Interatomic Potentials)," *Annalen der Physik*, **64**, 253-87 (1921).

16 B.G. Dick and A.W. Overhauser, "Theory of the Dielectric Constants of Alkali Halide Crystals," *Phys. Rev.*, **112**, 90-103 (1958).

17 R. Najafabadi, H.Y. Wang, D.J. Srolovitz, and R. LeSar, "A New Method for the Simulation of Alloys: Application to Interfacial Segregation," *Acta metall.*, **39**, 3071-82 (1991).

18 F.F. Abraham, N.-H. Tsai, and G.M. Pound, "Bond and Strain Energy Effects in Surface Segregation: an Atomic Calculation," *Surf. Sci.*, **83**, 406-22 (1979).

19 W.D. Kingery, H.K. Bowen, and D.R. Uhlmann, *Introduction to Ceramics*, 2nd Ed., p. 58. Wiley, New York, 1976.

Structure and Phase Equilibria

DEFECTS AND METASTABLE STRUCTURES OF $MgAl_2O_4$

S. P. Chen, M. Yan,
Theoretical Division, Los Alamos National Laboratory,
Los Alamos, NM 87545, USA

R. W. Grimes
Department of Materials, Imperial College, London, UK

J. D. Gale
Department of Chemistry, Imperial College, London, UK

INTRODUCTION

$MgAl_2O_4$, which crystallizes in the spinel structure, is usually found to be very resistant to radiation induced swelling and transformation at both room and elevated temperatures. Nevertheless recently, Yu et al [1,2] have found that under low temperature conditions (100K) a single crystal of $MgAl_2O_4$ spinel can be amorphized through irradiation by Xe atoms via an intermediate metastable phase. This intermediate phase has a higher Young's modulus, while the final amorphous phase has a lower Young's modulus than the original spinel structure [2]. The diffraction pattern of the metastable structure shows a reduction of the lattice parameter by approximately a factor of two, while the cubic symmetry along the [001] zone is maintained. Several cubic structures have been proposed for this metastable phase of $MgAl_2O_4$ [1, 3, 4].

In this paper, we present calculated results of the properties of both the normal spinel and inverse spinel structures of this material, as well as properties of point defects in the spinel structure. These results provide important information which can be used to further study and understand possible metastable states of this material. The calculated properties of the "amorphous" structure are also presented. In addition, we have calculated properties of several proposed metastable states of $MgAl_2O_4$ and the results can be found in [5]. The interionic interactions are described by Coulombic long range and the Buckingham short range potentials which incorporate the shell model [6] to account for ionic polarization of ions. The short range potentials were taken from [7]. The structural relaxations were performed by using the program GULP [8].

RESULTS

Fig.1 shows the normal spinel structure of $MgAl_2O_4$. If we consider the O^{2-} ions to form a f.c.c. sublattice, the Mg^{2+} ions occupy one out of eight tetrahedral interstitial sites, and the Al^{3+} ions two out of four octahedral interstitial sites within the O^{2-} f.c.c. unit cell. In an alternate structure, inverse spinel, the Mg^{2+} ions and half of the Al^{3+} ions are on octahedral interstitial sites, and the remaining half of the Al^{3+} ions are on the tetrahedral interstitial sites.

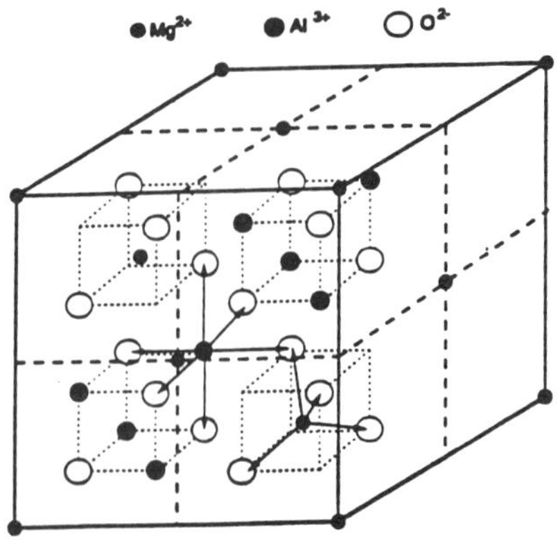

Fig.1 The structure of the spinel.

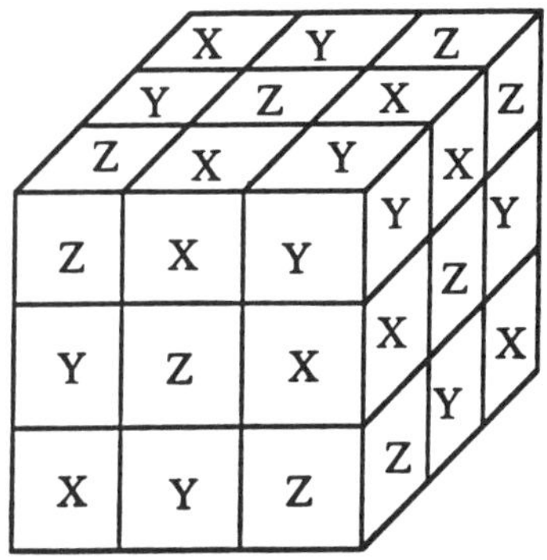

Fig.2. The building block of the "amorphous" state.

Table 1 summarizes the calculated equilibrium properties of $MgAl_2O_4$ in the normal and inverse spinel structures, in the proposed "rock salt" structure, and in the "amorphous" structure. They include the lattice energy, E_L, lattice volume V, lattice constants a, b, and c, structural symmetry, and the Young's modulus, E. For the spinel structure, the calculated lattice constant 8.1439Å is very close to the experimental measured value of 8.08Å, and the cubic symmetry is retained after relaxation. The internal variable x, which represents the position of O^{2-} ion, was calculated to be 0.387, which is the same as the experimental value. The Young's modulus, calculated as an average value based on the Voigt and Reuss assumptions [9], is 328 GPa for $MgAl_2O_4$ in normal spinel. Since the experimental value is 277 GPa, the calculated value is overestimated by about 18%. This is because in the fitting of the potentials the accuracies in energies and crystal structures are emphasized at the expense of the accuracy of the moduli. Therefore, the values of the moduli can only be used as a guide instead of as absolute values. As for the inverse spinel, the calculations show a higher lattice energy than for the normal spinel, indicating its structure is less favorable. The relaxed structure of the inverse spinel has almost tetragonal symmetry, with one of the angles equals 88.59° while the other two equal 90.00°.

Table 2 summarizes the calculated energies of several point defects in the $MgAl_2O_4$ spinel structure. It can be seen that the site exchange between an Mg^{2+} ion and an Al^{3+} leads to the smallest energy increase, 0.92eV/atom, while all other point defects lead to much larger energy increases. This suggests that the most likely point defects to be formed under irradiation are the disorder between cations, with metastable structures being formed subsequently. Among all the Frenkel defects, the Mg^{2+} ion has the smallest Frenkel defect energy, which is 4.59eV/atom. Even though, this energy is larger than the energy penalty of creating a Schottky defect, consisting of one vacancy on an Mg site, two vacancies on Al sites and four vacancies on O sites. This energy increase is 3.71eV/atom. It suggests that in $MgAl_2O_4$ the formation of Frenkel defects is more difficult than the formation of Schottky defects and more dramatically than the exchange of cations. These results are similar to those found for $ZnCr_2O_4$ spinel [10].

Table 1. Calculated properties for several structures of $MgAl_2O_4$.

Structures	E_L (eV)	V (Å^3)	a,b,c (Å)	symmetry	E (GPa)
normal spinel	-200.51	67.52	8.1439	cubic	328
inverse spinel	-199.22	66.32	8.1122 8.1122 8.0652	near tetragonal	352
"rock salt"	-193.03	63.40	4.1199 4.1199 3.7354	tetragonal	420
"amorphous"	-200.09	88.56	---------	triclinic	152

Table 2. Calculated energies of several point defects in $MgAl_2O_4$ spinel structure.

Defect	Energy (eV)	Defect	Energy (eV)
Mg'_{Al}	30.40	$Al^{\bullet}_{Mg}$	-28.56
$V''_{Mg}+Mg^{\bullet\bullet}_i$	9.18	$V'''_{Al}+Al^{\bullet\bullet\bullet}_i$	12.74
$V^{\bullet\bullet}_O+O''_i$	10.57	$V''_{Mg}+2V'''_{Al}+4V^{\bullet\bullet}_O$	25.94

The proposed "rock salt" structure of $MgAl_2O_4$ is the one in which the four O^{2-} ions form an f.c.c. unit cell, and the Mg^{2+} ion and two Al^{3+} ions are distributed randomly on the four octahedral interstitial sites within the f.c.c. unit cell. The calculated properties of one of the variants in this structure are summarized in Table 1. This structure has less binding energy, a smaller volume, and higher Young's modulus than those in the more usual spinel structures. The structure possesses tetragonal symmetry after relaxation, and the direction of the c-axis depends on how the Mg^{2+} ion and two Al^{3+} ions are distributed among the four octahedral interstitial sites. An "amorphous" state of $MgAl_2O_4$ was created by building up a structure with the unit cell consisting of 27 mixed "rock salt" structural units, as shown in the schematic plot in Fig.2. The small units denoted as x, y, or z represent the "rock salt" unit cells with the c axis pointing in the x, y or z directions, respectively. The unit cell for the "amorphous" state was built in the way such that each unit is neighbored by other units with different orientations, e.g., each x unit is neighbored by y and z units. The equilibrium properties of the relaxed "amorphous" state are given in Table 1. The lattice energy increases by only 0.2%, and the volume increases by 31% compared with the spinel structure. This small increase in the binding energy is not significant and it indicates that when the spinel is heavily irradiated it is energetically favorable to transform to the amorphous phase. Also, the expansion of the "amorphous" phase of 31% is consistent in sign with the irradiation experiment but no quantitative information is presently available. Furthermore, the calculated Young's modulus of the "amorphous" structure is smaller than that of the spinel structure by 54%, which agrees qualitatively with results from indentation experiments [2,5].

Figs.3(a-c) show, respectively, the coordination environments within 4Å around the Mg^{2+}, Al^{3+} and O^{2-} ions in the normal spinel structure. The Mg^{2+} and Al^{3+} ions are coordinated by four and six O^{2-} ions at the distance 1.93Å and 1.94Å, while the O^{2-} ion is coordinated by one Mg^{2+}, and three Al^{3+} ions. Fig.4 shows the average coordination numbers around each of Mg^{2+}, Al^{3+} and O^{2-} ions in the "amorphous" state of $MgAl_2O_4$, where X-Y represents the number of Y neighbors around each X atom. The most significant feature in the amorphous structure is the wide range of values for all the bond lengths between ions. For example, around each Mg^{2+} ion the O^{2-} neighbors are distributed in the range from 1.76Å to 2.24Å, and similarly for the Al^{3+} ion and O^{2-} ions. Such a wide range of bond lengths indicates that the diffraction pattern should show glass-like or amorphous-like halos instead of the sharp crystalline diffraction spots or rings. This is consistent with the experimental observations [1]. Table 2 summarizes the average bond length between ions in the "amorphous" state. It can be seen that on average, the Mg-O, O-O and Al-Al bond lengths become larger, and Al-O and Mg-Al bond lengths become smaller than those in the spinel structure.

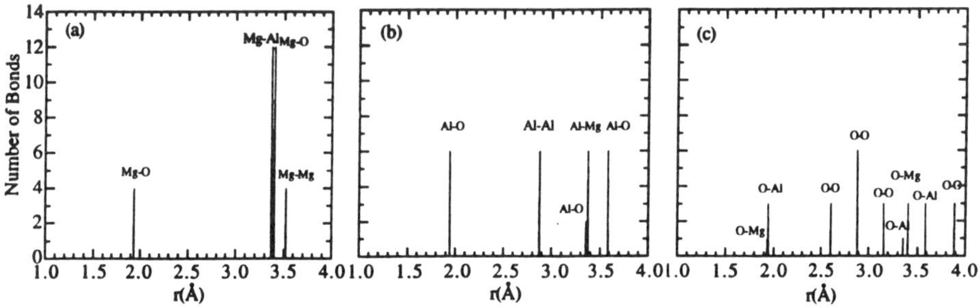

Fig.3. Number of bonds around (a) Mg^{2+} ions, (b) Al^{3+} ions and (c) O^{2-} ions in $MgAl_2O_4$ spinel structure.

Table 2. The average bond length between ions in the normal spinel and the "amorphous" state of $MgAl_2O_4$.

	Mg-O	Al-O	O-O	Mg-Al	Al-Al
normal spinel	1.93 Å	1.94 Å	2.89 Å	3.37 Å	2.89 Å
"amorphous"	2.00 Å	1.79 Å	3.00 Å	3.06 Å	3.25 Å

CONCLUSIONS

We predict, from atomistic simulations, that in the $MgAl_2O_4$ spinel structure the exchange of an Mg^{2+} ion with an Al^{3+} ion has the lowest energy penalty,

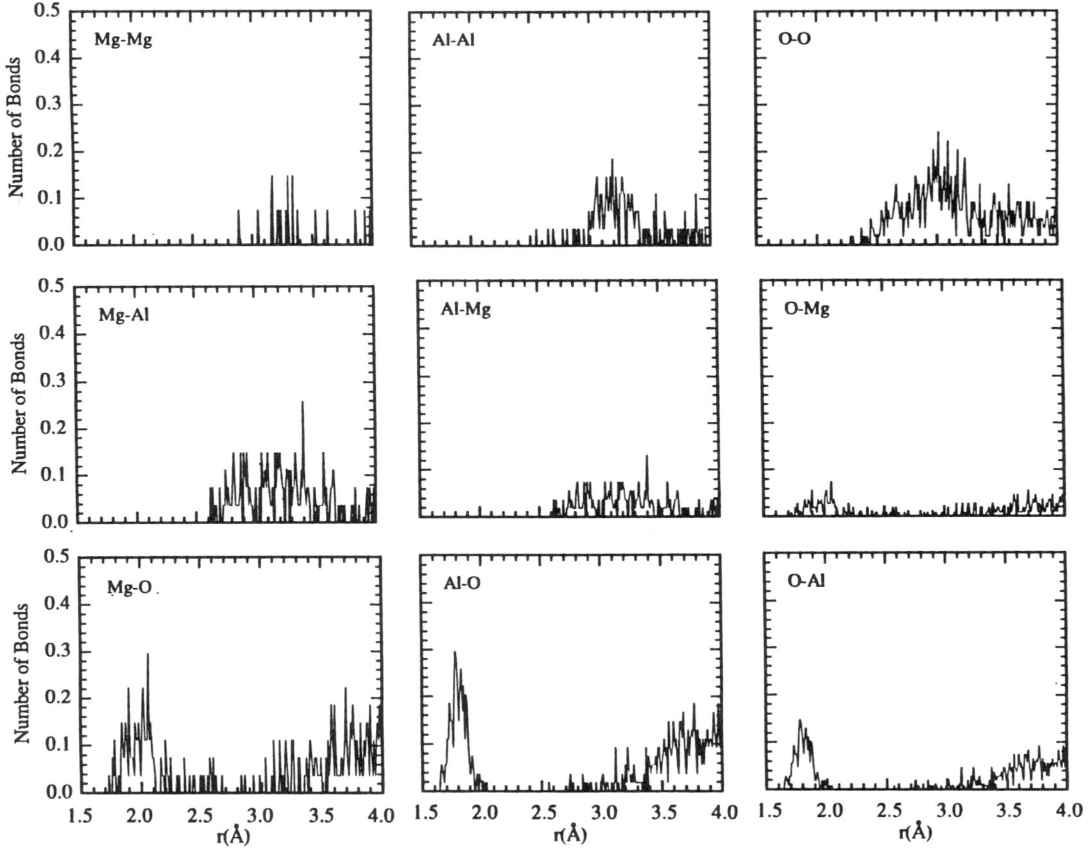

Fig.4. Average number of bonds around ions in the "amorphous" state of $MgAl_2O_4$ within the distance of 4Å

0.92eV/atom. The Schottky defect increases the energy by 3.71eV/atom. The formation of Frenkel defects is the most difficult, and increases the energy by at least 4.59eV/atom for the Mg^{2+} Frenkel defect. The proposed "rock salt" structure of $MgAl_2O_4$ has a tetragonal symmetry with a smaller volume and larger Young's modulus, and the "amorphous" state has a larger volume and smaller Young's modulus than those of the $MgAl_2O_4$ spinel structure.

ACKNOWLEDGMENT

We would like to thank Drs. K. E. Sickafus, R. Devanathan, N. Yu and M. Nastasi for help discussions. This work was supported by the Division of the Materials Science, Office of the Basic Energy Sciences of U. S. Department of Energy.

REFERENCES

1. N. Yu, K. E. Sickafus and M. Nastasi, *First Observation of Amorphization in Single-Crystal $MgAl_2O_4$ Spinel*, Philo. Mag. Lett. 70, 235 (1994).
2. N. Yu, K. E. Sickafus and M. Nastasi, *Temperature Effects on Ion Irradiation Damage in $MgAl_2O_4$ Spinel Single Crystals,* in Microstructure of Irrated Materials, (edited by I. M. Robertson, S. J. Zinkle, L. E. Rehn and W. J. Phythian), Mat. Res. Soc. Proc. Symp., Vol. 373, (1995).
3. L. M. Wang, W. L. Gong, N. Bordes, R. C. Ewing and Y. Fei, in Microstructure of Irrated Materials (edited by I. M. Robertson, S. J. Zinkle, L. E. Rehn and W. J. Phythian), Mat. Res. Soc. Proc. Symp., Vol. 373, (1995).
4. R. Devanathan, N. Yu, K. E. Sickafus and M. Nastasi, *The Structure of the Metastable State in Ion-Irradiated $MgAl_2O_4$ Spinel*, submitted to Philo. Mag. Lett. (1995).
5. S. P. Chen, M. Yan, J. Gale, R. W. Grimes, R. Devanathan, K. E. Sickafus, N. Yu and M. Nastasi, *Atomistic Study of Defects, Metastable and "Amorphous" Structure of $MgAl_2O_4$,* in preparation for Philo. Mag. A. (1995).
6. B. G. Dick and A. W. Overhauser, *Theory of the Dielectric Constants of Alkali Halide Crystals*, Phys. Rev. 112, 90 (1958).
7. T. S. Bush, J. D. Gale, C. R. A. Catlow and P. D. Battle, *Self-consistent Interatomic Potentials for the Simulation of Binary and Ternary Oxides*, J. Mater. Chem. 4, 1 (1994).
8. J. D. Gale, GULP-General Utility Lattice Program, Royal Institution of Great Britain, (1994).
9. D. H. Chung and W. R. Buessem, *The Elastic Anisotropy of Crystals,* in Anisotropy in Single Crystal Refractory Compounds (edited by F. W. Vahldick and S. A. Mersol), Plenum Press, (1968).
10. R. W. Grimes, J. D. Binks, A. B. Lidiard, "The solution of ZnO in $ZnCr_2O_4$ spinel", Phil. Mag. (1995) in press.

COMPUTER SIMULATION OF THE INFLUENCE OF BORON ON THE STRUCTURE OF BC_x.

Charles T. Hach, Susan E. Hill, and Linda E. Jones
NYS College of Ceramics at Alfred University
Alfred, New York 14802

ABSTRACT

The intent of this project is to determine the influence of boron on the structure of the novel compound BC_x. This material is a graphite-like film which is produced via a CVD reaction involving BCl_3 and C_6H_6. It has been determined that BC_x is more oxidation resistant than traditional carbon fibers. The ability of BC_x films to resist oxidation appears to be related to its structure. Earlier computational studies of BC_x involved calculations of the electronic structure of stoichiometric BC_3. In these studies, boron was considered only as a substitutional species. In the present study, MOPAC 7 will be used to produce a quick estimate of the structure and basic properties for various forms of BC_x. MOPAC 7 is a semi-empirical molecular orbital package. The host lattice can be modeled as a very large molecule. Truncated bonds are saturated with hydrogen atoms. Host and substitutional defect structures can be adequately simulated with a single layer.

INTRODUCTION

Carbon/carbon (C-C) composites are of great importance to the aerospace industry as a thermostructural material. C-C thermostructural materials were initially developed in 1958. Research in this area was supported early on by the U.S. Air Force Dyna-Soar space plane program and the NASA Apollo projects . With the advent of the space shuttle program came intensified research in the area of C/C composites **[1]**. The driving force behind studying structural carbon composites, is the need for materials which have higher specific strengths than conventional aluminum, titanium and nickel based alloys, and can withstand higher use temperatures **[2]**.

Because of their superior mechanical properties, carbon fiber is used as the reinforcement phase in C-C composites. These composites have in turn been used as a structural material in rocket nozzles, race car brakes, and in the nose cap and leading edges of the space shuttle. C-C composites are also a candidate material for the mechanically attached thermal protection system of the proposed National Aerospace Plane (NASP) **[3]**. The problem with these materials is oxidation at temperatures in excess of 400°C. One possible solution to this problem is the use of BC_x as an oxidation resistant alternative to carbon-carbon structural materials. The oxidation rates of BC_x coated carbon fibers are less than as - received carbon fibers.

BC_x is a material with a graphite-like structure containing up to 17% boron in the lattice. This novel compound was first produced in the mid 1980's **[4]**. The reaction of interest for the synthesis of BC_x is the vapor deposition reaction at 700°C to 1000°C between BCl_3 and benzene to produce a solid boron-carbon film and hydrogen chloride gas. Under these conditions, a Friedel Crafts-like reaction occurs **[5]**. Benzene rings are linked by boron atoms depositing a hexagonal BC_x structure (Figure 1). The focus of the present work is to model the influence of boron on the BC_x structure.

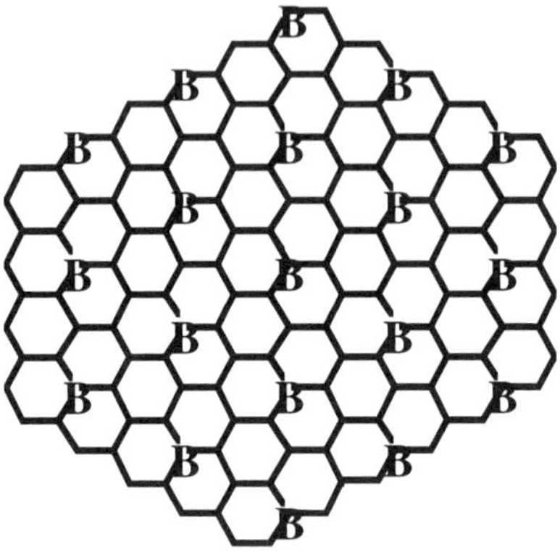

Figure 1 - Schematic representation of the proposed BC_x structure.

MOPAC 7 is a semi - empirical molecular orbital package. The Semi - empirical Hamiltonian AM1 (Austin Model 1) is used in the electronic part of the calculation to obtain molecular orbitals, heat of formation, and its derivative with respect to molecular geometry. Molecular orbital parameters for AM1 are optimized so as to reproduce the experimental heat of formation to form a mole of compound at 25°C from its elements in their standard state **[6]**. The formalism used in AM1 is essentially the same as in MNDO (Modified Neglect of Diatomic Overlap) with the exception of the CRF (Core Repulsion Function) **[7]**.

In previous studies BC_x was modeled as a periodic lattice of stoichiometric composition BC_3 **[8,9]**. In both studies the structure and electronic structure of BC_3 was investigated. The calculated B-C bond length was found to be ~0.155 nm and the C-C bond length ~0.142 nm. When BC_3 is modeled as a single layer it is calculated to behave as an indirect band gap semi - conductor. However, when BC_3 was modeled as a multi-layered structure it was predicted to behave like a metal which is supported by its metallic appearance **[8,4]**.

COMPUTATIONAL METHOD

All calculations were carried out using the IBM SP2 Supercomputer located at the Cornell Theory Center (CTC) (Cornell University, Ithaca, New York). The program MOPAC 7 was used for all calculations.

It is well established that graphite consists of layers of closely bound carbon atoms, while the layers themselves are relatively isolated from one another **[10]**. BC_x has been shown experimentally to have a layered structure similar to that of graphite **[4,5]**. The structure being modeled in the present study is of a single BC_x layer. This layer is modeled as a finite cluster with hydrogen at the edges. Three different host lattices were studied: Benzene (C_6H_6), Coronene ($C_{24}H_{12}$), and Circumcoronene ($C_{54}H_{18}$). In each case the host lattice (or molecule) was modeled and used as a "baseline" structure. Then the influence of substituting boron for carbon on selected sites within the host molecule (or cluster) was determined. Finally, evidence of convergence for the model was summarized.

All computer generated images were generated by using the program Insight II. Cartesian coordinates for the optimized structures were downloaded into program Insight II from the MOPAC 7 output file.

RESULTS AND DISCUSSION

The heats of formation of the host molecules calculated in the present study are compared to calculated results obtained by Wang and Frenklach in Table I **[11]**. In both cases the program MOPAC 7 was used while utilizing the semi - empirical Hamiltonian AM1. It is apparent that the results obtained in the current study are

Table I - Comparison of current results to previously calculated heats of formation of various hydrocarbon host molecules.

Molecule	HOF (kcal/mol) AM1[a]	HOF (kcal/mol) AM1[b]
Benzene (C_6H_6)	22.0	22
Coronene ($C_{24}H_{12}$)	96.2	96.2
Circumcoronene ($C_{54}H_{18}$)	226.9	226.9

[a]Wang and Frenklach [11]
[b]Present study

similar to those calculated in the previous work. For reference, the experimental value for the heat of formation of Benzene at 25°C is ~ 20 kcal/mol **[12]**. The results given in Table I provide a measure of confidence for the use of the previously described computational methodology to study the influence of boron on the structure of BC_x.

Figure 2 is a computer generated image of a) a benzene host molecule (C_6H_6) and b) a benzene host molecule with boron substituted into the structure at two carbon sites ($B_2C_4H_6$). In comparing the geometry of the two structures it is apparent that the presence of substitutional boron results in distortion of the original host structure. Similar results can be seen in the larger molecules as a result of similar substitution orientations as can be seen in Figure 3. Figure 3 is a computer generated image of a) a Circumcoronene molecule ($C_{54}H_{18}$) and b) a circumcoronene molecule with boron substituted onto two carbon sites

($B_2C_{52}H_{18}$). The distortion in a molecules structure as a result of boron substitution can most easily be observed by overlapping the "wireframe" representation of the structure of the host molecule with that of a molecule that contains boron substitutionally, as can be seen in Figure 4. It can be clearly seen that the presence of boron in the host molecule results in distortion of that molecule.

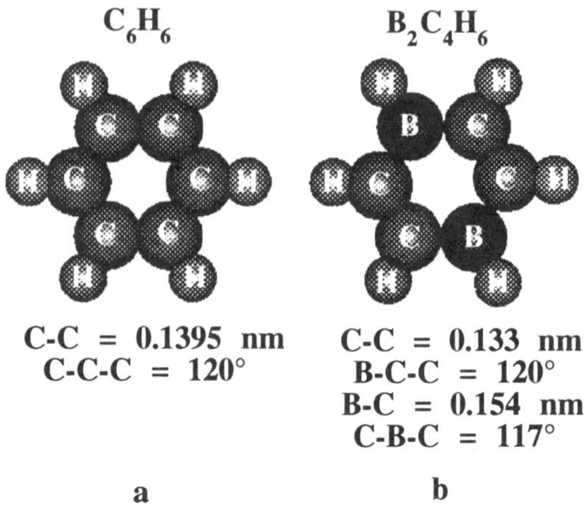

Figure 2 - Computer generated structure of a) Benzene host molecule, b) Benzene host molecule with boron substituted onto two carbon sites.

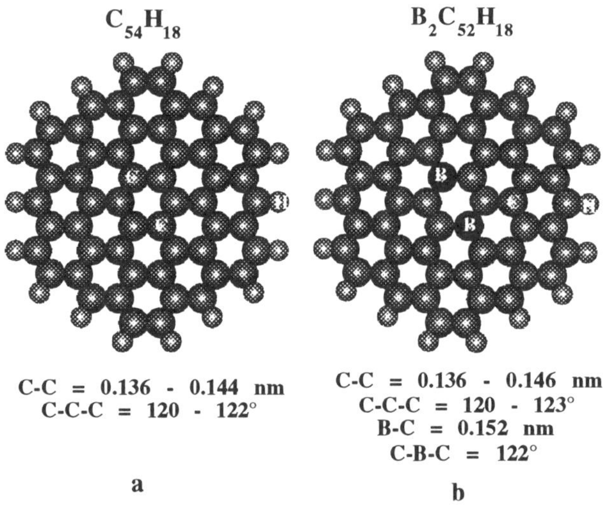

Figure 3 - Computer generated structure of a) Circumcoronene ($C_{54}H_{18}$) host molecule, b) Circumcoronene host molecule with boron substituted onto two carbon sites ($B_2C_{54}H_{18}$).

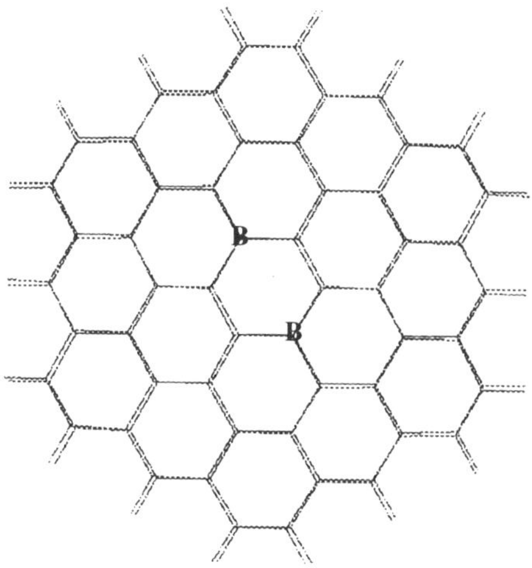

Figure 4 - Computer generated wireframe representation of the structure of a $C_{54}H_{18}$ molecule overlapped with the structure of $B_2C_{54}H_{18}$.

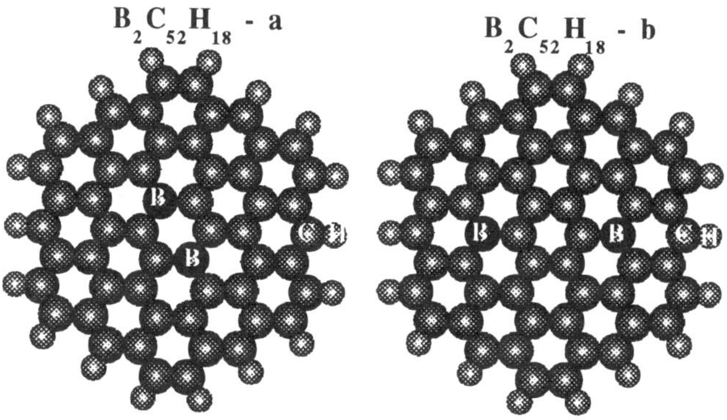

HOF = 278 kcal	**HOF = 325 kcal**
C-C = 0.136 nm	**C-C = 0.136 - 0.146 nm**
B-C = 0.152 nm	**B-C = 0.151 nm**
C-C-C = 120 - 123°	**C-C-C = 117 - 124°**
C-B-C = 122°	**C-B-C = 119 - 120°**

Figure 5 - Computer generated images of circumcoronene with boron substituted for two carbon atoms in two different orientations.

A second point of interest is the influence of different defect orientations on the structure of the host molecule. Two different boron substitution orientations in the circumcoronene ($C_{54}H_{18}$) host molecule are shown in Figure 5. Figure 5a is a computer generated image of circumcoronene with boron substituted for two carbon atoms in the central carbon ring. Figure 5b is a similar image of circumcoronene with boron substituted for two carbon atoms on two different outer carbon rings. By studying the geometrical values (i.e. C-C bond lengths) it appears that the two structures are similar with minor perturbations. However, when their heats of formation are compared it is apparent that there is a marked difference in the boron - boron interaction from one defect orientation to another.

In order to validate the current approach for modeling the structure of BC_x, evidence of convergence in the proposed model was sought. The calculated heats of formation for all three host molecules studied are given in Table II. In addition the calculated heats of formation for the host molecules with boron substituted for carbon on two sites in the central carbon ring are tabulated. It is apparent that as the molecule size is increased the percent change in HOF decreased. Such a decrease is evidence of convergence.

Table II - Comparison of the calculated heats of formation of host molecules to molecules with boron substituted for two carbon atoms in the central carbon ring.

Molecule	HOF (kcal)	2B substituted HOF (kcal)	ΔHOF (kcal)	ΔHOF (%)
C_6H_6	22	63	41	186
$C_{24}H_{12}$	96	176	80	83
$C_{54}H_{18}$	227	278	51	22

Secondly the total energies of the host molecules were compared to the total energies of the boron substituted molecules, as shown in Table III. In this case the percent change in total energy decreased as the cluster (molecule) size was increased. This is further evidence of convergence in the proposed model. Finally, the electronic charge on the substituted boron was studied as a function of molecular size. The electronic charge on the substituted boron is plotted as a function of molecular size in Figure 6. It can be seen that the three points can be fit with a logarithmic function indicating that the charge on boron is converging rather than diverging as cluster size is increased.

Table III - Comparison of the calculated total energies of host molecules to molecules with boron substituted for two carbon atoms in the central carbon ring.

Molecule	Total Energy (eV)	B substituted Total Energy (eV)	ΔTotal Energy (eV)	ΔTotal Energy (%)
C_6H_6	-850	-731	119	14
$C_{24}H_{12}$	-3237	-3116	121	3.7
$C_{54}H_{18}$	-7160	-7041	119	1.7

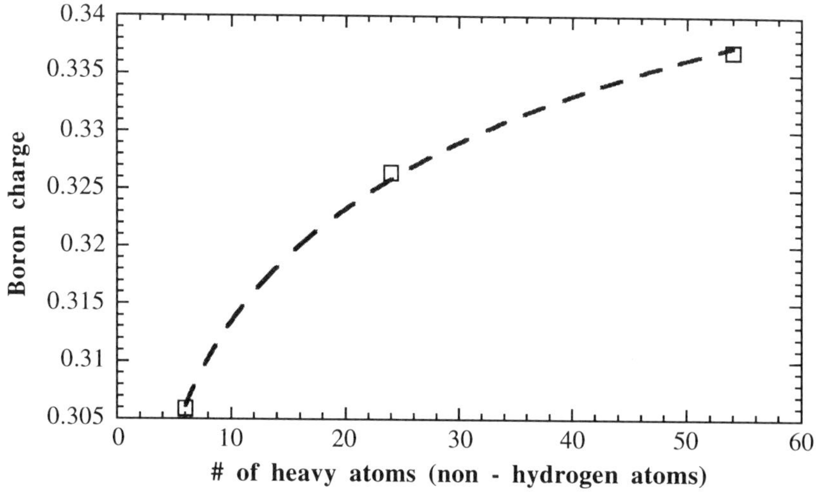

Figure 6 - Electronic charge on boron atoms as a function of the number of heavy atoms (non - hydrogen atoms) in a particular molecule.

CONCLUSIONS

The cluster size convergence with respect to boron substitution heats of formation, total energies, and charges is satisfactory. There are marked differences in the boron - boron interaction for different defect orientations.

FUTURE WORK

In future studies larger cluster sizes will be investigated. Within these larger cluster sizes different defect configurations and orientations will be studied. Finally, ab initio techniques (i.e. DMOL) will be used to model interlayer interactions, and interstitial impurities.

REFERENCES

1. J.D. Buckley, "Carbon-Carbon , An Overview," Ceramic Bulletin, 67 [2] 364-368 (1988).

2. J. A. Dicarlo, "High Temperature Structural Fibers-Status and Needs," NASA Technical Memorandum 105174, (1991).

3. T.M.F. Ronald, "NASP Launches New Materials," Advanced Materials and Processes, 144 [3] 24-27 (1993).

4. J. Kouvetakis, R.B. Kaner, M.L. Sattler, and N. Bartlett, "A Novel Graphite-Like Material of Composition BC_3, and Nitrogen-Carbon Graphites," J. Chem. Soc., Chem. Commun., 1758-1759 (1986).

5. D.L. Fecko, L.E. Jones, P.A. Thrower, "The Formation and Oxidation of BC_3, A New Graphitelike Material," Carbon, 31 [4] 637-644 (1993).

6. J.J.P. Stewart, Mopac Manual (1990).

7. M.J.S. Dewar, E.G. Zoebisch, E.F. Healy, and J.J.P. Stewart, "AM1: A New General Purpose Quantum Mechanical Molecular Model," J. Am. Chem. Soc., 107 3902-3909 (1985).

8. D. Tomanek, R.M. Wentzcovitch, S.G. Louie, and M.L. Cohen, "Calculation of Electronic and Structural Properties of BC_3," Physical Review B, 37 [6] 3134-3136 (1988).

9. Y. Lee and M. Kertesz, "The Electronic Structure of BC_3," J. Chem. Soc., Chem. Commun., 75 (1988).

10. R.O. Brennan, "The Interlayer Binding in Graphite," J. Chem. Phys., 20 [1] 40-48 (1952).

11. H. Wang and M. Frenklach, "Enthalpies of Formation of Benzenoid Aromatic Molecules and Radicals," J. Phys. Chem., 97 3867-3874 (1993).

12. G.M. Barrow, *Physical Chemistry, 4th Edition,* McGraw - Hill p. 803 (1979).

Stress Design of Ceramic Grain Boundary

Jing kun Guo
Shanghai Institute of Ceramics, Chinese Academy of Sciences
1295 Dingxi Rd., Shanghai 200050, China

1.Introduction

The development of ceramic materials has trended to various forms of multiphase composites, exactly called as multiphase composed ceramic[1], which content:fibre(or whisker) enhanced ceramic composites, the second phase particulate dispersed multiphase composed ceramics, self-enhanced multiphase composed ceramics, gradient multiphase composed ceramics and etc.. Hence, it is important on the bonding among different phase to some properties of multiphase composed composites. The author has ever suggested to clarify the different between "interface" and "interphase"[2]. With regard to the former, only its physical matches have to be considered; but for the later, it is more important to consider the chemical compatibility between two phases than the physical matches. Hence, it was suggested that the grain boundary area was considered as an independent phase, which was made to fit similarly to three principles for ceramic material design[3]. In this paper, stress design of ceramic grain boundary, based on the interface and interphase studies of multiphase composed ceramics is put forward. It is expected that, through stress design of boundary, ceramic composites materials possess the ability of transferring, consuming or absorbing the applied loading energy so as to achieve the ceramic strengthening and toughening, even to solve the problem of ceramic brittleness.

2.The actions of grain boundary stresses

According to the behaviors of stress, the stresses existing in the material may be beneficial to the mechanical performance, especially to the strength, but sometimes may be harmful to.

The idea of this paper is to extend the stress state existing in the materials to the grain boundary area of ceramic materials. The applied stress and the interior stresses must interact at the boundary. As a result, the relationship of main crack with stress area at the grain boundary as shown in Fig. 1:

(1).the consumption of the main crack energy in Fig. 1(a)

(2).the stresses at the tip of main crack weaken or blunt when it meets the tensile stress or shear stress during its propagating, and then the microcrack is formed rapidly in the grain boundary area

(3).the branch of a main crack is formed due to the formation of microcrack

(4).the deflection of main crack occurs due to the formation of microcrack

(5).the pinning effect will hinder the main crack to expand continuously

(6).the grain boundary debonds, and then consumes the main crack energy.

the results as above mentioned will increase the fracture energy, which contributes not only to the enhancement of fracture toughness of materials, but also to solution of the problem of ceramic brittleness intrinsically due to the consumptive ability to

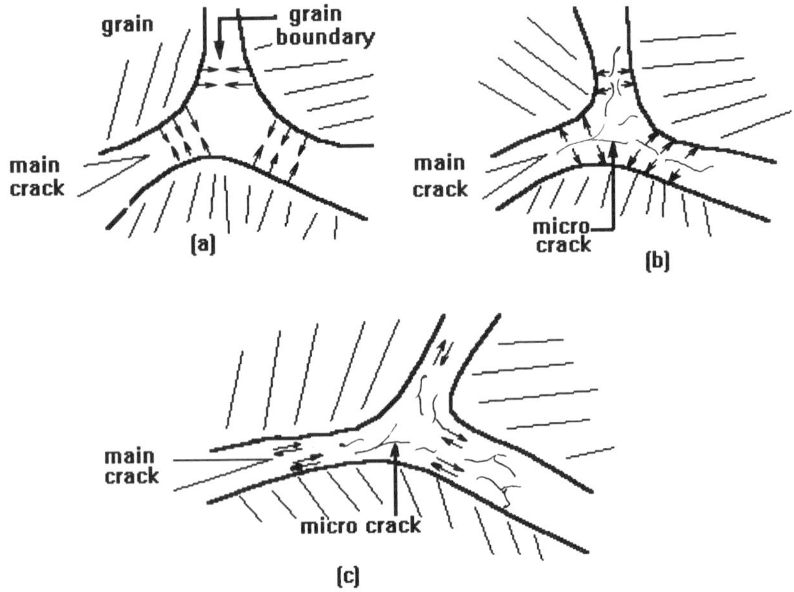

Fig.1 The relation of main crack with stress area at grain boundary.
(a).The stress area in compressive stress state.
(b).The stress area in tensile stress state, micro-crack formation.
(c).The stress area in shear state, micro-crack formation.

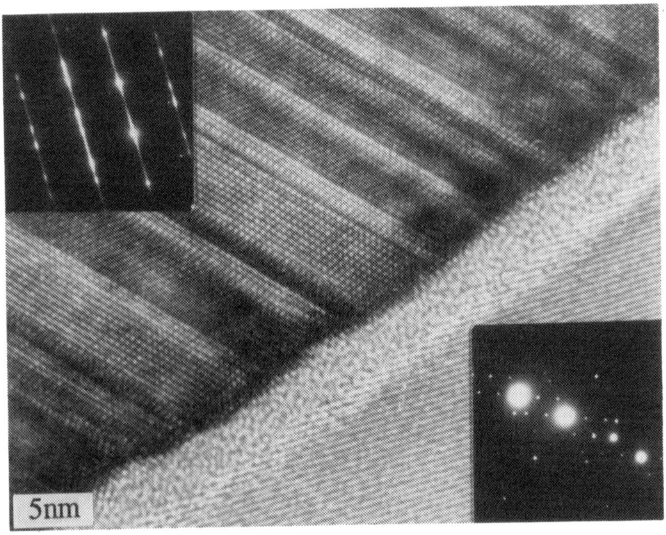

Fig.2 The grain boundary phase of SiC(w)/Si_3N_4[4]

the applied energy. Hence, the stress design of grain boundary is hopeful as an effective approach for the solution of ceramic brittleness completely.

3. Stress design of grain boundary

The grain boundary phase, as an independent phase, is the key point for the reasons that an interface and interphase must be considered separatively in the ceramic multiphase composed composites. It coexists with main crystal phase, not subordinate, and connects with main crystal phase. It is also an important part for the ceramic material design. In the experiment, at first, the behavior and character of grain boundary phase should be studied, and then the relations with the main crystal phase. These relations and the effects of environment will result in the formation of the different stress states at the grain boundary phase area. Therefore, we are able to adjust the intrinsic character of the grain boundary phase, the effects by environment, and the relation with the main crystal phase in order to attain the aim of the stress design of grain boundary.

The various stress effects of grain boundary are shown in Fig. 2-5.

Table 1 and 2 show the different strengthening and toughening effects, for example, particulate dispersion, ZrO_2 phase transformation and whisker enhancement[5,6]. When two strengthening and toughening mechanisms are acting together(No.4), something the overlap effect appears which indicates simple addition of each mechanism(No.2 and No.3), but sometimes there are a concordant effect with an additional effect(as shown in Table 1)

viz. $\Delta\sigma_4 > \Delta\sigma_2 + \Delta\sigma_3$, $\Delta K_{IC-4} > \Delta K_{IC-2} + \Delta K_{IC-3}$

the additional effects will be studied deeply in coming days. These consequences offer the best information and theoretical base the optimization on stress design of grain boundary, and then the best strengthening and toughening effects for ceramic materials will be achieved by the stress design of grain.

4.Summary

Extending the absorbable or consumable applied energy mechanism existing inside the ceramic materials to the grain boundary area creates the possibility to solve the ceramic brittleness. When some strengthening and toughening mechanisms take a role together in multiphase composed ceramics, there exist not only the simple addition of energy mechanism, but also the concordant effect with addition effect. It offers an effective approach for the stress design of grain boundary of ceramic materials.

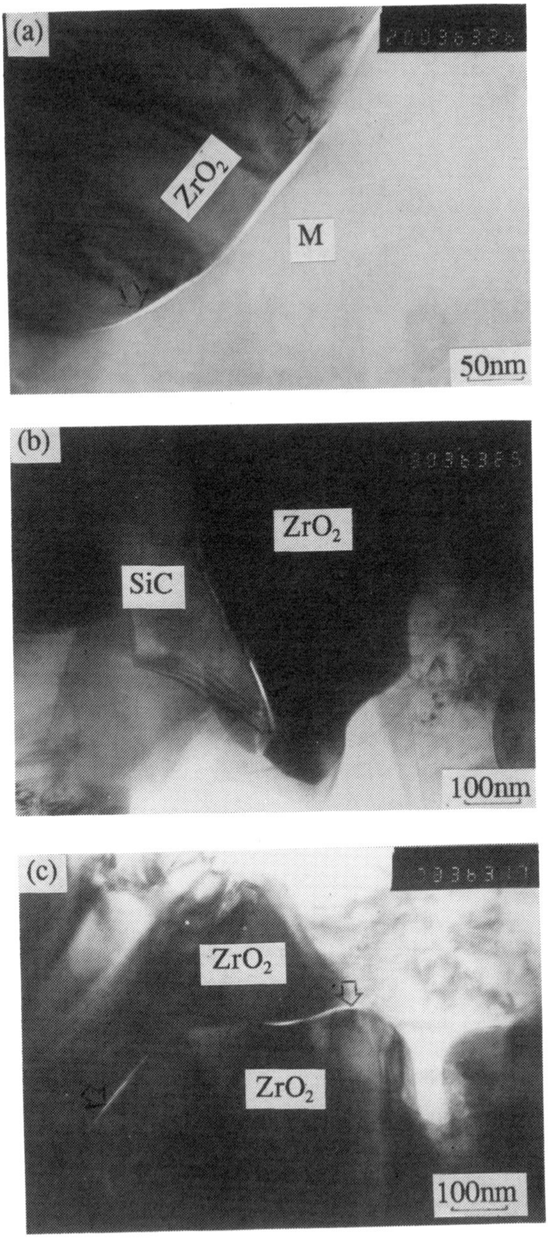

Fig.3 The interphase (a and b) and interface in SiC(p)/Y-TZP/Mullite multiphase composed ceramics[5]

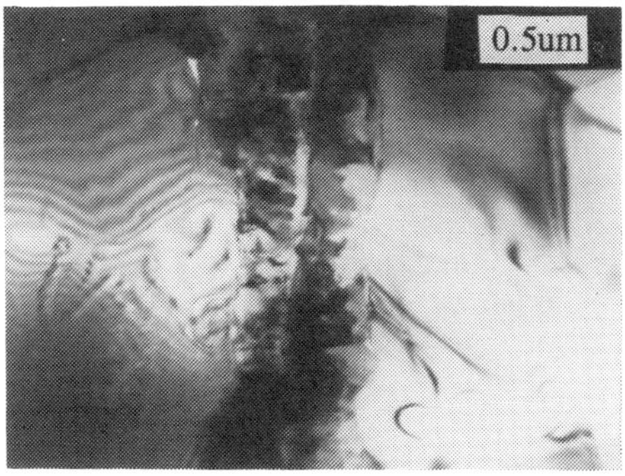

Fig.4 The stress stripes and dislocations appear in SiC(w)/AlN[6]

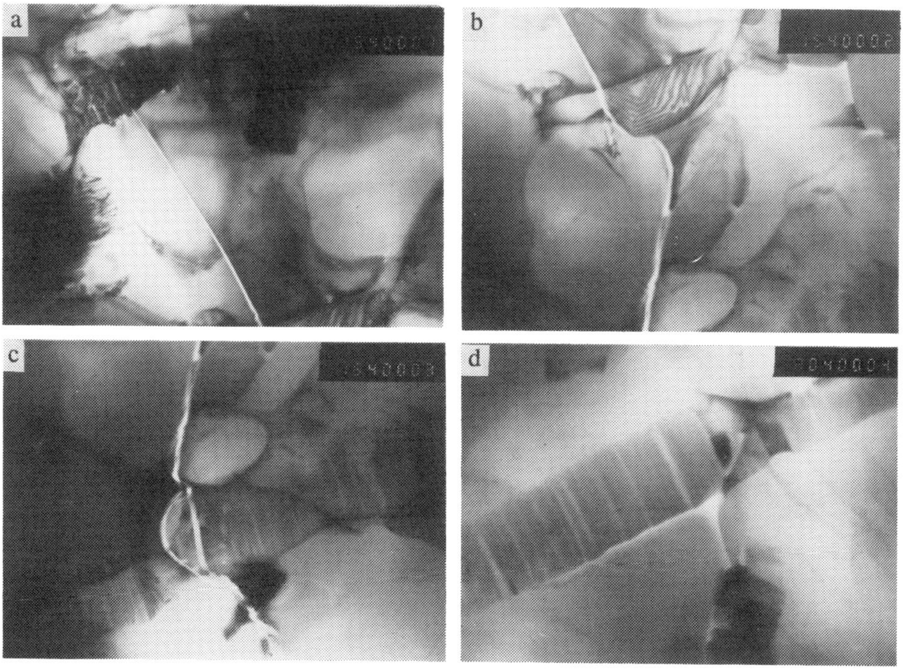

Fig.5 (a).Crack pinning (b).Crack deflection
(c).Crack branching (d).Grain boundary debonding in SiC(w)/AlN[6]

Reference

1.Jingkun Guo, J.Solid Chem.,96,No.1(1992)108

2.Jingkun Guo, Litai Ma, "Study on the Interface and Interphase of Ceramics", 5th Inter. Symp. on Ceramic Materials and Components for Engines, Shanghai, China, May 29-June 1,(1994).

3.Jingkun Guo, Peinan Zhu, Proc. of Inter. Conf. on High Temperature Ceramic Matrix Composites, Bordour, France, 20-24, Sep.(1993).

4.Yufeng Zhang, Xiaoxian Huang, Jingkun Guo, etc. J.Inorganic Mater. 7,No.3,329,(1992).

5.Xiaoxian Huang, Jinsheng Hong, Jingkun Guo, etc. Proc. of 4th Inter. Symp. on Ceramic materials and Components for Engines, edited by R.Carlsson, T.Johnsson and L.Kahlman. Elserier Aralieil Science.(1992).

6.Xin Jiang, Report of PHD, Shanghai Institute of Ceramics, Chinese Academy of Sciences (1994).

Table 1 The strength and fracture toughness of Mullite matrix multiphase Ceramics[5]

No	Materials	σ/MPa	$\Delta\sigma$/MPa	K_{IC}/MPa.m$^{1/2}$	K_{IC}/MPa.m$^{1/2}$
1	Mullite		246	2.9	
2	20V/O Y-TZP/Mullite	397	151	4.9	2.0
3	30V/O SiC(p)/Mullite	357	111	4.6	1.7
4	20V/O Y-TXP/30V/O SiC(p)/Mullite	610	364	6.7	3.8

Table 2 The strength and fracture toughness of AlN matrix multiphase Ceramics[6]

No	Materials	σ/MPa	$\Delta\sigma$/MPa	K_{IC}/MPa.m$^{1/2}$	K_{IC}/MPa.m$^{1/2}$
1	AlN	308		3.12	
2	10W/O TiC(p)/AlN	342	34	4.05	0.93
3	20W/O SiC(w)/AlN	520	212	5.22	2.10
4	10W/O TiC(p)/20W/O SiC(w)/AlN	566	258	6.76	3.64

X-Ray Absorption Near Edge Spectra and Electronic Structure of $La_{2-x}(AE)_xCuO_{4-y}$ (AE=Ca, Sr and Ba)

Y. Wu, D.E. Ellis and S.A. Hoffman
Northwestern University, Evanston IL 60208

Abstract

Theoretical and experimental X-ray absorption near edge spectra (XANES) in alkaline earth (AE) doped La_2CuO_{4-y} are presented. Experimental measurements have been made on the Ca K- and Ba L_3-edges, which are also studied in the framework of density functional theory. Charge densities and potentials are calculated by the self-consistent-field embedded cluster approach, final states were determined by the multiple-scattering theory, and the absorption cross-section was calculated in the dipole approximation. Charge distributions, orbital populations and densities of states are also reported, and correlated with XANES data.

INTRODUCTION

The electrical and thermal conductivities of cuprates are highly correlated with their defect structures, with particular interest in the La_2CuO_4 system focusing upon excess oxygen interstitials and the divalent impurity-vacancy complexes which may be optimized with respect to superconducting properties. Recent experimental studies on $La_{2-x}AE_xCuO_{4-y}$ were made over a range of doping levels ($0 < x < 0.16$), oxygen partial pressures ($10^{-5} < P(O_2) < 1$ atm), and temperatures ($700 < T < 1000$ C). This work shows that thermoelectric powers and electrical conductivities are similar for Ca-, Sr- and Ba- doped materials whereas the oxygen non-stoichiometric behavior differs notably for calcium doping. Jonkers plot analysis indicates holes as the major charge carriers, and that the hole concentration increases with doping level. The data are consistent with band-type conduction; however, existing defect structure models involving either isolated or associated oxygen vacancies were found inadequate to reconcile electrical properties and observed oxygen non-stoichiometry[1-5].

In order to gain further insight for the defect structures one needs a combination of local spectroscopic probes and theoretical analysis. We have carried out XANES measurements on the Ca K-edge and Ba L_3-edge in well characterized samples, at the National Synchrotron Light Source. Using the well-developed Discrete Variational method, we have employed Density Functional theory to calculate and interpret related electronic structure parameters. The embedded cluster approach, including a variational expansion of the cluster orbitals in atomic orbital (LCAO) form, permits extraction of local atomic configurations, energy levels, and charge and spin densities[6,7]. Self-consistent potentials were obtained for clusters embedded in the infinite solid; these potentials were projected onto muffin-tin form in order to carry out multiple-scattering analysis of the continuum final states. The calculated dipole-absorption spectra are then compared directly with experimental XANES. This work complements the previous oxygen K-edge studies of Fischer et al.[8]

DEFECT MODELS

We have considered two electrically neutral defect complexes AE_2O_V consisting of two adjacent divalent impurities and a nearby oxygen vacancy, consistent with the transport data. The first complex, P1, places O_V in the CuO planes; the second complex, P2, places O_V on the axial position. A third complex, P3, consisting of CaO_V was also considered as a candidate for its unusual properties, but will not be discussed here for lack of space.

A typical variational cluster consists of 43 atoms embedded in the infinite ideal host crystal; we briefly summarize the AE effective atomic configuration and charges in Table 1.

Table I. Effective atomic configuration and charges of alkaline earth impurities

Site	Ion Charge	Configuration
P1	Ca 1.92	$3p^{5.96}3d^{0.14}$
	Sr 1.94	$4p^{5.92}4d^{0.13}5s^{0.02}$
	Ba 1.98	$5p^{5.81}5d^{0.20}6s^{0.01}$
P2	Ca 1.95	$3p^{5.97}3d^{0.12}$
	Sr 1.90	$4p^{5.92}4d^{0.12}5s^{0.06}$
	Ba 1.95	$5p^{5.81}5d^{0.19}6s^{0.05}$

We can see that in both sites, the substituents are found close to the nominal +2 ionic charge, with small amounts of excited state character mixed in due to metal-oxygen interaction. As reported in previous cluster studies, the effective ionic charges of La, Cu, and O are somewhat less than the formal valence values, falling in the range La: +2.6-2.9e, Cu: +1.5-1.9e, and O: -1.7 to -2.0e depending upon the local environment. The calculated upper valence band is ~7-8 eV in width, consisting primarily of O 2p and Cu 3d structures (recall the nominal O^{2-} $2p^6$, Cu^{2+} $3d^9 4s^0$ configurations). The lower valence band, starting ~12 eV below the Fermi energy E_F, is due mostly to La 5p and O 2s character. The impurity valence levels, Ca 3p, Sr 4p, and Ba 5p span the lower valence band region. The unoccupied or conduction band states form the important final states of the XANES spectra, and will be discussed below. As an example, we display the total density of states of clusters representing Ba_2O_V P1 and P2 sites in Fig. 1.

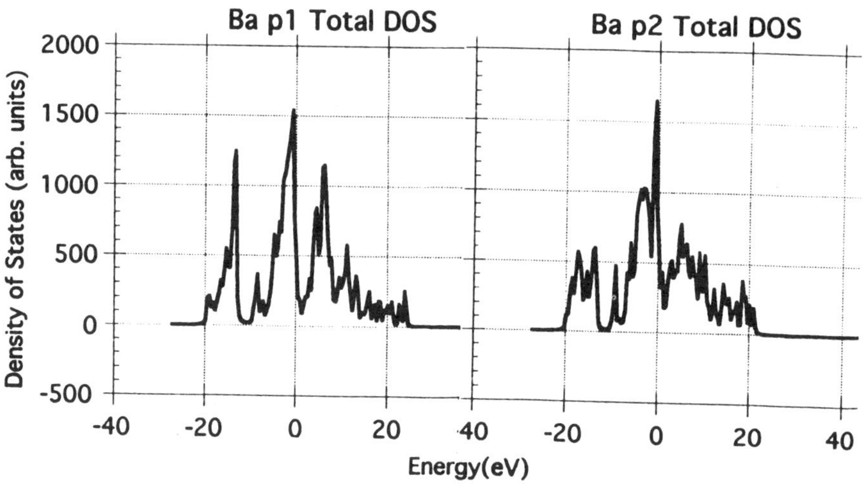

Fig. 1. Total density of states for 43 atom Ba_2O_V defect clusters. The energy zero is the Fermi energy.

X-RAY ABSORPTION SPECTRA

The experimental Ca K-edge spectra (shown in Fig. 2) are dominated by the white-line transition into the Ca 4p resonance. In addition one sees several pre-edge features, and undulating structure due to backscattering from neighbors at higher energies which merges with the EXAFS oscillations. The multiple scattering cross-section shows similar features, and semiquantitative agreement with experiment, once the muffin-tin zero of energy is chosen to align the 4p peak.

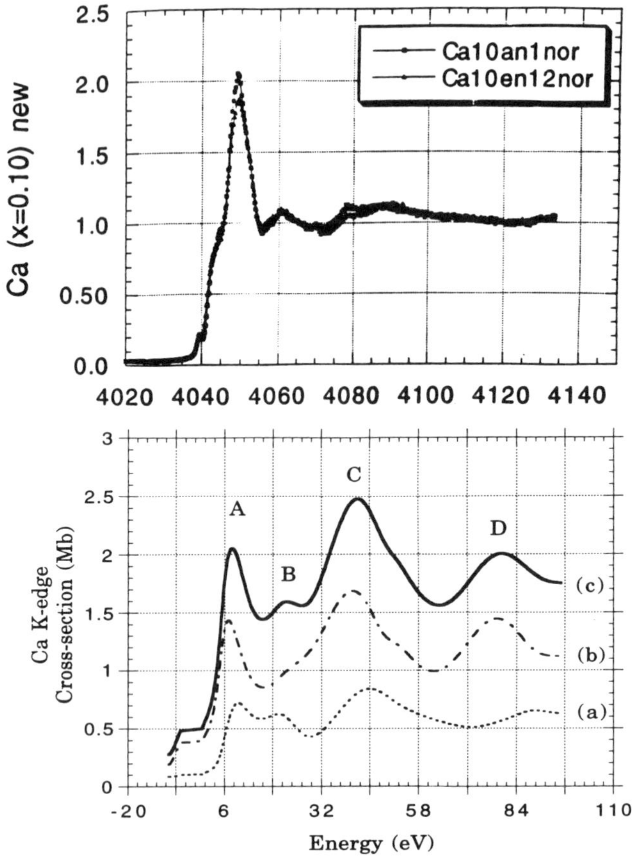

Fig.2. Experimental(upper) and theoretical (P1 site) (lower) Ca K-edge XANES for $La_{2-x}Ca_xCuO_{4-y}$ for x=0.10

A detailed analysis will be presented elsewhere; here we note that the defect geometry is not very well suited to the demands of the muffin-tin approximation, even making use of the standard 10-15% atomic sphere overlaps. This is doubtless responsible for some of the discrepancies between theory and experiment; in addition, we expect that local lattice relaxation not included in this work would shift the position of ligand-related peaks and also change the width of the 4p resonance. This structural sensitivity can be best exploited in a model using the full electronic potential. We can get some further insight for the nature of the XANES final states by considering the DV projected densities of states for $E > E_F$, given in Fig. 3.

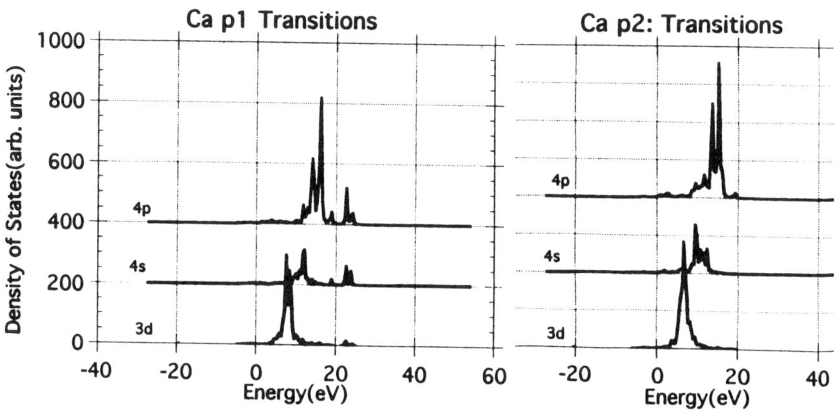

Fig.3. Conduction band projected density of states of Ca 3d, 4s, 4p character for Ca_2O_V at P1 and P2 sites.

From this figure we can see that the prominent pre-edge features are associated with Ca 3d states which, although dipole-forbidden at the atomic level, are made accessible by the low local symmetry and mixing with oxygen hole states. Similar structure has been seen in a variety of low symmetry transition metal oxides. Interestingly, there is no similar pre-edge feature in the experimental Ba L_3 spectra (Fig. 4) even though the calculated Ba 5d band does lie in this energy region.

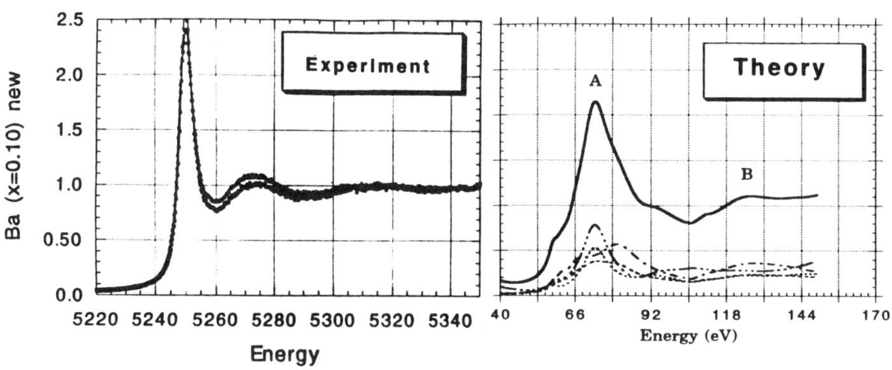

Fig.4. Experimental and theoretical (P1 site) Ba L_3-edge XANES for $La_{2-x}Ba_xCuO_{4-y}$ for x=0.10

CONCLUSION

Measured Ca K-edge and Ba L3-edge spectra of impurity complexes in $La_{2-x}AE_xCuO_{4-y}$ reveal sufficient structure to be of utility in assigning the local defect cluster composition and geometry. The present density functional embedded cluster calculations give a preliminary view of the underlying electronic structure and its site dependence. Limitations of the standard muffin-tin approximation used in multiple-scattering models of absorption have been noted; it is suggested that use of the full electronic potential will be useful in making more quantitative comparisons with experiment. The present model results are in semiquantitative agreement with experiment; local lattice relaxation should be considered next in making a definite site assignment.

ACKNOWLEDGMENTS

Research supported by the U.S. Dept. of Energy, Grant No. DE-FG02-84ER45097. Access to computer facilities at the National Energy Research Supercomputing Center is gratefully acknowledged.

REFERENCES

1. J.M. Longo and P.M. Raccah, "The Structure of La_2CuO_4," J. Solid State Chem. **6** 526 (1973).

2. T.O. Mason, "Defect Chemistry of High T_c Superconductor Cuprates", in "Ceramic Materials for Electronics", J. Nowotny, ed. (1991).

3. M.-Y. Su, E.A. Cooper, C.E. Elsbernd and T.O. Mason, "High-Temperature Defect Structure of Lanthanum Cuprate," J. Am. Ceram. Soc., **73** 3453 (1990).

4. L Shen and T.O. Mason, "High Temperature Electrical Properties and Defect Chemistry of $La_{2-x}Ca_xCuO_{4-y}$ Superconductors (I): Experimental," submitted; L Shen and P.A. Salvador and T.O Mason, "High Temperature Electrical Properties and Defect Chemistry of $La_{2-x}Ca_xCuO_{4-y}$ Superconductors (II): Defect Structure Modeling," submitted.

5. H. Shaked, J.D. Jorgenson, B.A. Hunter, R.L. Hitterman, K. Kinoshita, F. Izumi and T. Kamiyama, "Defect Structure and Superconducting Properties of $La_{1.8}Sr_xCa_{1.2-x}Cu_2O_{6-\delta}$," Phys. Rev. **B48**, 12941(1993).

6. D.E. Ellis , J. Guo, and D.J. Lam, "Embedded Cluster Models of Ceramic Electronic Properties," Reviews of Solid State Science **5** 287 (1991)

7. H. Chen and J. Callaway, "Electronic Structure of Copper-Oxygen Clusters in the High-Tc Superconductor $La_{2-x}Sr_xCuO_4$," Phys. Rev. **B40**, 8800 (1989); H. Chen and J. Callaway, "Local Electronic-Structure and Magnetism of 3D Transition-Metal Impurities (Cr, Mn, Fe, Co, and Ni) in $La_{2-x}Sr_xCuO_4$," Phys. Rev. **B44**,2289(1992).

8. D.A. Fischer, A.R. Moodenbaugh and Youwen Xu, "Oxygen K near-edge X-ray Absorption Spectroscopy of $La_{2-x}M_xCuO_4$(M=Ca, Sr and Ba)," Physica **C215**, 279(1993).

First-Principles Computation of the CaO-MgO Phase Diagram

G. Ceder, P. D. Tepesch, A. F. Kohan and G. D. Garbulsky
Department of Materials Science and Engineering
Massachusetts Institute of Technology, Cambridge MA 02139

C. Coley, H. T. Stokes, L. L. Boyer and M. J. Mehl
Naval Research Laboratory, Complex Systems Theory Branch, Washington DC 20375

B. P. Burton
National Institute of Standards and Technology, Gaithersburg, MD 20899

1. Introduction

Considerable interest exists in the development of models to predict structures and phase diagrams in materials systems. Much emphasis has been on metallic systems where the combination of first-principles band structure methods and statistical thermodynamics [1,2,3] has proven to be a powerful tool to compute binary and ternary phase diagrams with little or no experimental input. For these systems it has been found that the computation of the structural formation energies remains the largest source of error in the final result. In this paper, we use the cluster expansion technique [3,5] to compute the phase diagram of the CaO-MgO system and compare the results for different energy methods. We find that the phase diagram can be predicted by using accurate quantum mechanical methods.

2. Configurational Energy Dependence: The Cluster Expansion

CaO and MgO both form the rocksalt structure and have substantial solubility in each other. In CaO-MgO mixtures, Ca and Mg ions are distributed over the fcc cation lattice while the oxygen sublattice remains fully occupied. The different ionic radius of Mg and Ca causes local deformations of the rocksalt lattice so that the ions do not necessarily occupy the positions of an ideal lattice. These relaxations are fully incorporated in the model we will describe. Our only assumption is that a one-to-one correspondence between lattice sites and ions remains.

To obtain the free energy of a CaO-MgO mixture one needs, in principle, the energy of all possible configurations of Mg and Ca on the cation sublattice. The type of ion associated with site i on the cation lattice can be labeled with an occupation variable, σ_i, which takes the value of +1 (-1) for a Mg (Ca) ion. To obtain the energy, E, as an explicit function of the cation configuration, $\{\sigma\}$, E is expanded in an orthogonal set of basis functions [5]:

$$E(\{\sigma\}) = \sum_{\alpha} V_{\alpha} \sigma_{\alpha} \tag{1}$$

where α is any cluster of points on the lattice. The summation in eqn. (1) is over all possible combinations of lattice points (clusters). The functions σ_α are defined as the product of all occupation variables σ_i in cluster α and form a complete and orthogonal set of basis functions in the space of all configurations. Eqn. (1) is therefore formally exact. The expansion coefficients V_α are called Effective Cluster Interactions (ECI) and their value can be determined by fitting eqn.(1) to empirical or first-principles energy computations. These effective cluster interactions have to be distinguished from potentials. The ECI are numbers and have no distance-dependence. They parameterize the dependence of the energy on the configuration of cations, and incorporate changes in lattice parameter and ionic relaxations.

The free energy of the system can be computed by integrating eqn. (1) numerically or analytically (approximately) with standard lattice model techniques. In this model, only the configurational entropy is included in the free energy. The effect of vibrations can be incorporated by expanding, in eqn. (1), the free energy of the system with fixed cation configuration instead of the static ground state energy. In the latter case, the ECI are temperature dependent. In the harmonic approximation this temperature-dependence can be linearized above the Debye temperature [6]. We will therefore write the ECI as a chemical part and a vibrational part:

$$V_{\alpha}(T) = V_{\alpha}^{chem} + k_B T V_{\alpha}^{vib} \tag{2}.$$

The configuration-independent term in the vibrational free energy is excluded from eqn. (2). More details of this procedure can be found in ref [6]

To determine the values of the chemical ECI we have computed the zero-temperature energy of structures with a number of different cation arrangements and fitted the temperature-independent part of the ECI (first term in eqn. (2)) to them. More energies are used than unknown ECI so that a least-square fit can be performed. We included pair interactions up to sixth nearest neighbor cation-cation distance and several multiplet interactions. Convergence of the expansion was tested by systematically leaving structures out of the fit and testing how their energy is predicted by the cluster expansion

To evaluate the accuracy of different energy approximations we computed the direct energies both with empirical potential models [7] and with the Self-Consistent Potential Induced Breathing method (SCPIB) [8,9], also referred to as the Spherical Self-Consistent Atomic Deformation (SSCAD) method. The SCPIB is a first-principles method in which the total energy of the crystal is computed as the sum of the self-energy of each ion plus the interatomic interaction energy. The interaction term is computed from the overlap of the self-consistent ionic charge densities. For the empirical energy model we used a Buckingham pair potential augmented with a shell model to treat the polarizability of each ion. Many parameter sets for this model have been published for the MgO-CaO system[7,10,11].

Figure 1 shows results for the formation energy for 21 structures with different cation arrangements. The largest structure contained 64 ions in the unit cell. The energy of all structures was computed with the SCPIB method (open circles) and with the potential parameters taken from Lewis and Catlow [7]. The formation energy for a structure with composition $Ca_{1-x}Mg_xO$ is defined as the difference between the total energy of the structure and the total energy of the pure oxides. We plot formation energy rather than the total cohesive energy as it is the relevant quantity for phase stability. The energy of all structures is minimized with respect to cell parameters as well as internal degrees of freedom (ionic positions). The formation energies in fig. 1 are all positive in agreement with the experimentally observed miscibility gap in this system. The chemical ECI derived from these formation energies are shown in figure 2: SCPIB (empty circles), empirical potentials (filled circles). The cluster expansion used in the fit contains pair interactions up to the sixth cation-cation nearest neighbor, three three-body terms and two four-body terms.

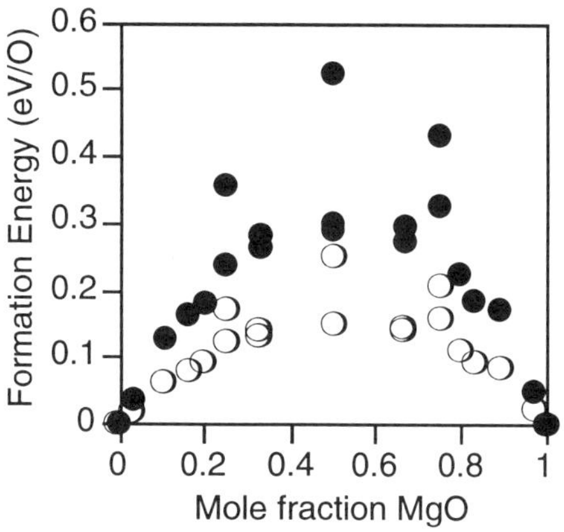

Fig. 1: Formation energies (eV per molecule) for structures with different ordered cation arrangements. Filled circles are obtained with the empirical potential model. The open circles are obtained with the SCPIB method.

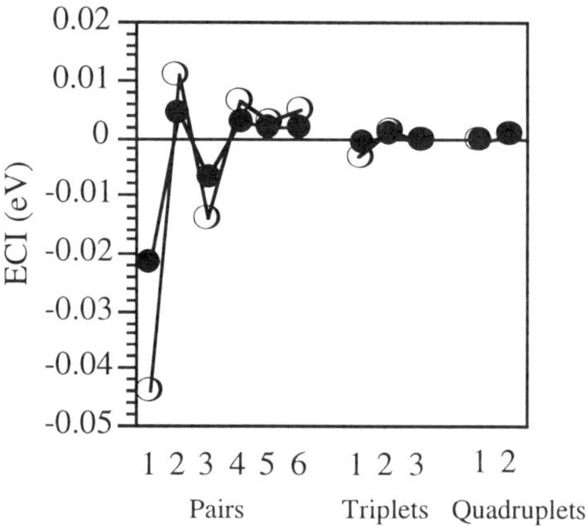

Fig. 2: Values of the effective cluster interactions obtained from the data in figure 1 for the first six nearest neighbor pairs, three triplets and two quadruplets.

The vibrational components to the ECI, V_α^{vib}, are given in Table I. They were determined in a similar procedure, by fitting them to the vibrational free energy of structures with a number of different cation arrangements. The vibrational free energy was computed by diagonalizing the dynamical matrix for each structure at a large number of k-points. The elements of the dynamical matrix were obtained indirectly from the SCPIB wave functions [4]. Note from eqn. (2) that the vibrational ECI are dimensionless.

Table I: Vibrational ECI for the CaO-MgO system

constant term	point term	nearest neighbor pair	nearest neighbor triplet
17.942045	0.308093	0.006327	0.005073

3. THE PHASE DIAGRAM

The total free energy of the system was computed by applying a combination of Monte Carlo simulation and the Cluster Variations Method (CVM) [1,5,12,13]. Monte Carlo simulations were used to obtain the internal energy and the probability distributions for cation arrangements on 13 and 14 point clusters [14]. The latter were then used in the CVM to compute the configurational entropy [15]. With this procedure free energies can be obtained in a single Monte Carlo run.

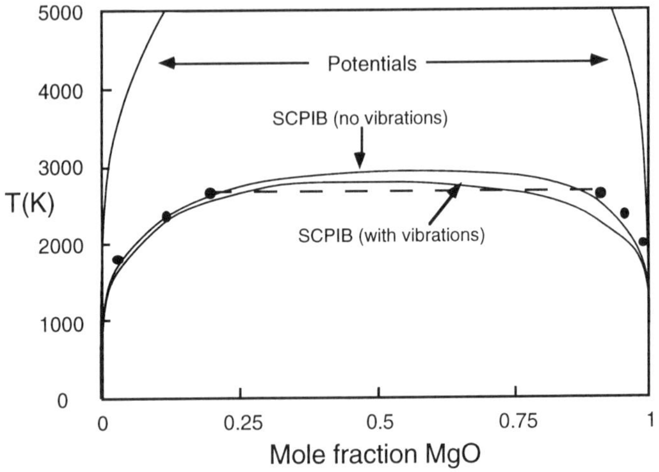

Fig. 3: Solid part of the CaO-MgO phase diagram. The top line were obtained with the Lewis and Catlow potentials [7]. The lower lines are the result of the SCPIB calculations (with and without vibrational effects). The filled circles and dashed line are respectively the experimental solubility limits and the eutectic temperature.

The resulting phase diagrams display a miscibility gap and are shown in figure 3. As only the solid state part is computed, the region above the experimentally observed eutectic temperature (dashed line) is metastable. Whereas the empirical potential parameters (upper line) significantly underestimate the solubility limits, the SCPIB (lower line) prediction is in excellent agreement with the experiment (solid circles). The effect of vibrations is to increase solubility limits. In this particular system the vibrational ECI are small. The maximum temperature of the miscibility gap is lowered by about 10%, reducing the agreement with the experimental data. The correction due to the vibrations is asymmetric: The increase in solubility of CaO in MgO is much greater than the increase in solubility of MgO in CaO.

4. DISCUSSION

The cluster expansion is clearly a versatile tool for the parameterization of the configurational energy dependence in this system. With only a few effective cluster interactions we can correctly parameterize the dependence of the energy on the cation configuration. One might expect that the rapid convergence of the cluster expansion is limited to systems such as CaO-MgO in which the ΔE_{mix} term has no long-range electrostatic term since the charges on the ions are the same. However, we have recently shown that the cluster expansion remains short-ranged even in alio-valent systems [16].

Considering the fact that no adjustable parameters were used in our phase diagram calculation, the agreement between the SCPIB prediction and the experimental solubility limits is extremely good. Errors of the order of 100% in transition temperatures are not uncommon in first-principles calculations for metallic systems. The fact that the SCPIB phase diagram without vibrations is somewhat closer to the experimental data is coincidental. Rather, these results indicate that the SCPIB slightly underpredicts the formation energies in this system. We verified this assumption by comparing the SCPIB formation energies for three structures with those obtained from highly accurate pseudopotential calculations [4]. From the results in table II it is clear that pseudopotential energies are higher than the SCPIB values. Although we did not pursue a full phase diagram calculation with the pseudopotential method, as it is extremely computing intensive, we expect it to give a very accurate phase diagram, given the above results.

Table II. Formation energies per molecule for three ordered structures

ΔE_f (eV/cation)	$L1_2$ (Ca_3MgO_4)	$L1_0$ ($CaMgO_2$)	$L1_2$ ($CaMg_3O_4$)
Empirical Potentials Lewis-Catlow	0.356	0.510	0.428
Pseudopotential	0.186	0.282	0.262
SCPIB	0.172	0.250	0.208

The empirical potential parameters published for the MgO-CaO system predict the qualitative features of the CaO-MgO system, including the sign of the formation energies and the asymmetry of the miscibility gap. However, given the highly ionic nature of this system, the lack of quantitative agreement using these parameters is somewhat unexpected. The pseudopotential and SCPIB calculations allow us to gain some insight into the energy effects responsible for this quantitative failure of the potential models. We used the pseudopotential wavefunctions to determine the charge density around an oxygen ion in the pure CaO and MgO structures. Figure 4 shows the charge inside a sphere centered around the oxygen ion as a function of the sphere radius. The dashed (solid) curve is for the oxygen in the CaO (MgO) structure. The oxygen ion in the CaO structure has more charge at a larger distance from the center of the ion, resulting in a somewhat larger ionic size. This result indicates that the size of the oxygen ion will be variable in mixtures of CaO and MgO and will depend on the types of cations around it: This is known as the oxygen "breathing" effect. In addition to this size effect, the self-energy of the oxygen ion is different in CaO and MgO (even at the same lattice constant). These multi-body effects cannot be captured well with a pair potential model. In recent work [4] we have found that when empirical potentials are fitted to experimental data such as elastic constants, the multi-body effects are forced into the two-body potential form and result in a potential that significantly overestimates the energy of mixing CaO and MgO. On the other hand, potentials fitted to closely reproduce the mixing quantities, result in a poor reproduction of other physical properties (elastic constants, dielectric constants, vibrational properties, etc.). Given the multi-body nature of the oxygen breathing and the limited number of parameters in the potential models, it seems unlikely that one set of parameters will simultaneously reproduce all of the relevant properties in these materials.

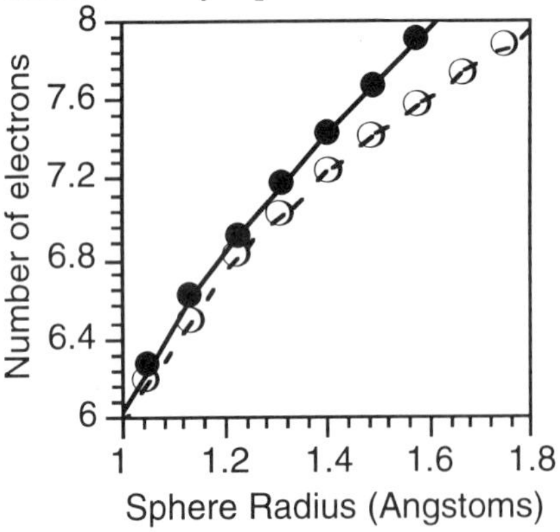

Fig. 4: Amount of valence charge in a sphere centered around an oxygen ion as a function of sphere radius in the MgO (solid curve) and CaO (dashed curve) structures.

5. CONCLUSION

Using a first-principles model, with no adjustable parameters, we have been able to compute the CaO-MgO phase diagram in excellent agreement with experimental data. This result is encouraging and indicates that accurate first-principles predictions for ionic systems are possible. Quantitative agreement with experimental data could not be achieved with previously published values of empirical pair potential parameters.

The technique of cluster expanding the configurational energy dependence is found to work well. The cluster expansion technique is not limited to systems with binary disorder, such as the cation disorder in the CaO-MgO system. We recently showed how a similar expansion can be used to study defect arrangements in alio-valent oxides with disorder on both the cation and anion sublattice [17], as well as in materials with multi-component disorder on a given sublattice [18]

ACKNOWLEDGMENTS5

The authors thank Dr. Julian Gale for giving us the GULP program to do the empirical potential calculations. The pseudopotential calculations were performed at the Pittsburgh Supercomputing Center.

6. REFERENCES

(1) de Fontaine, D. (1994). Cluster Approach to Order-Disorder Transformations in Alloys. In H. Ehrenreich & D. Turnbull (Eds.), Solid State Physics (pp. 33-176). Academic Press.

(2) Ceder, G. (1994). The ab initio Computation of Phase Diagrams. In R. J. B. D. Bloor M.C. Flemings and S. Mahajan (Eds.), Encyclopedia of Advanced Materials (pp. 1951-1956). Pergamon Press.

(3) Ceder, G. (1993). A Derivation of the Ising Model for the Computation of Phase Diagrams. Computational Materials Science, **1**(2), 144-150.

(4) Tepesch, P.D. et al. (1995). A model to compute phase diagrams in oxides with empirical or first-principles energy methods and application to the solubility limits in the CaO-MgO system. submitted to J. Am. Cer. Soc.

(5) Sanchez, J. M., Ducastelle, F., & Gratias, D. (1984). Generalized cluster description of multicomponent systems. Physica, **128A**, 334-350.

(6) Garbulsky, G. D., & Ceder, G. (1994). Effect of lattice vibrations on the ordering tendencies in substitutional binary alloys. Phys. Rev. B, **49**, 6327-6330.

(7) Lewis, G. V., & Catlow, C. R. A. (1985). Potential models for ionic oxides. Journal of Physics C: Solid State Physics, **18**, 1149-1161.

(8) Boyer, L. L., & Mehl, M. J. (1993). A self consistent atomic deformation model for total energy calculations: application to ferroelectrics. Ferroelectrics, **150**, 13-24.

(9) Boyer, L. L., Mehl, M. J., & Peterson, M. R. (1994). Charge transfer effects in $KNbO_3$ and sodium. Ferroelectrics, **151**, 7.

(10) Grimes, R. W. (1994). Solution of MgO, CaO and TiO_2 in alpha-alumina. J. Am. Cer. Soc., **77**, 378-384.

(11) Bush, T. S., Gale, J. D., Catlow, R. A., & Battle, P. D. (1994). Self-Consistent Interatomic Potentials for the Simulation of Binary and Ternary Oxides. Journal of Materials Chemistry, **4**, 831-837.

(12) Kikuchi, R. (1951). A theory of cooperative phenomena. Phys. Rev., **81**, 988-1003.

(13) Ducastelle, F. (1991). Order and Phase Stability in Alloys. Amsterdam: North-Holland.

(14) Tepesch, P. D., Ceder, G., Wolverton, C., & de Fontaine, D. (1992). Ab Initio Calculation of the Pd-V Fcc Superstructure Phase Diagram with Fourth Nearest Neighbor Interactions. In P. D. Bristowe, J. Broughton, & J. M. Newsam (Ed.), Materials Theory and Modelling, Proc.Mat.Res.Soc. Vol.291 (pp. 129-134).

(15) Schlijper, A. G., & Smit, B. (1989). Two-sided Bounds on the Free Energy from Local States in Monte Carlo Simulations. J. Stat. Phys., **56**, 247-260.

(16) Ceder, G., Garbulsky, G. D., & Tepesch, P. D. (1995). A convergent real-space expansion for configurational disorder in ionic systems. Phys. Rev. B, in press.

(17) Tepesch, P. D., Garbulsky, G. D., & Ceder, G. (1995). Model for the configurational thermodynamics in ionic systems. Phys. Rev. Lett, **74**, 2272-2275.

(18) Ceder, G., Garbulsky, G. D., Avis, D., & Fukuda, K. (1994). Ground states of a ternary lattice model with nearest- and next-nearest neighbor interactions. Phys. Rev. B, **49**, 1-7.

PHASE EQUILIBRIA AND DEFECT INTERACTIONS IN THE CUBIC $Gd_xZr_{1-x}O_{2-x/2}$ SYSTEM WITH EMPIRICAL POTENTIALS

Patrick D. Tepesch and Gerbrand Ceder
Massachusetts Institute of Technology
77 Massachusetts Avenue
Cambridge, MA 02139

INTRODUCTION

Lattice models are powerful tools for studying configurational thermodynamics. For many years these models have been successfully applied in many binary and ternary metallic systems to study phase equilibria with little or no experimental input[1]. Recently, the same models have been shown to work very well in ceramic systems including high temperature superconductors[2], and the MgO-CaO system[3]. Two developments have made the extension of these models to systems with aliovalent doping practical. First, aliovalent systems can be treated as coupled binary systems with the same techniques as simple binary systems[4]. Second, it has been shown that the conditionally convergent coulomb interactions can be well approximated with short-ranged effective interactions[5].

The $Gd_xZr_{1-x}O_{2-x/2}$ system is particularly well suited for the application of a lattice model. Except very near the pure compositions, the system is in the defect fluorite structure or, near x = 0.5, the related pyrochlore structure. In the defect fluorite structure the Gd and Zr ions are distributed on an fcc lattice with oxygen ions and vacancies distributed in the tetrahedral sites. The pyrochlore structure is a specific long-range ordered arrangement of the ions and vacancies on the fluorite lattice[6].

Experimental data are difficult to obtain and sparse, leaving an opportunity for modeling to provide useful insights. There has been some experimental work on phase equilibria in this system[7], but the limited amount of data does not fully characterize the nature of the pyrochlore to defect fluorite transition nor the short range order near the transition. Also, the ionic conductivity of doped ZrO_2 goes through a maximum as a function of M_2O_3 (e.g. M = Gd, Y) dopant concentration, however the detailed microscopic nature of this behavior is not well understood. Finally, it has been shown that at intermediate dopant concentrations the vacancies tend to sit near Zr ions[8], but it is not known how this changes with temperature and concentration. The models that have been developed to interpret the behavior at low concentrations in terms of defect interactions are based on various simplifying assumptions[9, 10], and are therefore applicable over a limited concentration and temperature range. In this paper we use a lattice model with which the pyrochlore --> defect fluorite transition and defect interactions at any concentration or temperature can be studied with no *a priori* assumptions.

MODEL

Each sublattice is represented by a binary lattice model[4]. On the anion sublattice, the occupation of each site is represented by a variable, σ, which takes the value 1 (-1) if an anion (vacancy) occupies the site. Similarly, on the cation sublattice, the occupations are represented by a variable δ, which takes the value 1 (-1) if a Zr (Gd) cation occupies the site. With these definitions, any configuration of the system can be expressed by two vectors,

$$\vec{\sigma},\vec{\delta} = \left\{\sigma_1,\sigma_2,\ldots,\sigma_{N_a}\right\},\left\{\delta_1,\delta_2,\ldots,\delta_{N_c}\right\} \tag{1}$$

where N_a and N_c are the number of anion and cation sites respectively. The ions are not fixed to the lattice sites; only a unique mapping of every arrangement of the ions to the sites of the lattice is necessary for the model to be applied.

With this representation of the system, the energy as a function of the arrangement of the ions on the lattice can be expressed as sums of products of the occupation variables on all the sites:

$$\begin{aligned} E\left(\vec{\sigma},\vec{\delta}\right) \equiv {} & \sum_i V_1^a \sigma_i + \sum_i V_1^c \delta_i + \sum_i \sum_j V_{i,j}^a \sigma_i \sigma_j + \sum_i \sum_j V_{i,j}^c \delta_i \delta_j + \sum_i \sum_j V_{i,j}^{a,c} \sigma_i \delta_j + \\ & \sum_i \sum_j \sum_k V_{i,j,k}^a \sigma_i \sigma_j \sigma_k + \sum_i \sum_j \sum_k V_{i,j,k}^c \delta_i \delta_j \delta_k + \sum_i \sum_j \sum_k V_{i,j,k}^{a,c} \delta_i \delta_j \sigma_k + \ldots + \ldots \end{aligned} \tag{2}$$

In this expression the coefficients, V, are called Effective Cluster Interactions (ECI). There are three types of ECI: Cation-cation (V^c), anion-anion (V^a), and anion-cation or coupling terms($V^{a,c}$). This is an exact expansion of the energy if all terms are included. Fortunately, for many systems, the expansion is rapidly convergent and only a few pair and multiplet terms are necessary to obtain an accurate energy expression. The values of the ECI for a system are usually obtained by truncating the expansion and fitting the non-zero ECI to the energies of many ordered arrangements of the ions computed using any total energy method. For this work, the energy of the $Gd_xZr_{1-x}O_{2-x/2}$ system is computed using pair potentials of the form:

$$V\left(r_{ij}\right) = A_{ij} \exp\left(-\frac{r_{ij}}{\rho_{ij}}\right) - \frac{C_{ij}}{r_{ij}^6} \tag{3}$$

with the coulombic interactions determined by the full ionic charges on each of the ions (Zr = 4, Gd = 3, O = -2). The parameters for the potentials were taken from the literature and are given in Table I.

Table I. Potential Parameters

Potential	A	ρ	C
Zr-O[11]	985.869	.37600	0.0
Gd-O[12]	1336.76	.35509	0.0
O-O[11]	22764.3	.14900	27.88

Using well known statistical mechanics methods such as Monte Carlo, the Cluster Variation Method[13] (CVM), or a low temperature expansion, the configurational entropy can be included to compute the properties of the system as a function of temperature and composition. For this work, the Monte Carlo method was used with 5184 ion sites and periodic boundary conditions to examine the pyrochlore to defect fluorite phase transition and the oxygen ion behavior at low concentration of Gd_2O_3. One powerful benefit of a lattice model is the detailed information about the microscopic behavior of the system. This will be shown here by plotting the relative occurrence of two particular ion arrangements on the 5-point cluster of sites made up of a cation tetrahedra with an anion site in the center. This data provides information about cation-vacancy association as a function of temperature and concentration.

RESULTS AND DISCUSSION

The energies of a large number of configurations (~100) are needed to determine the Effective Cluster Interactions (ECI). The energies of 1032 periodic structures with primitive unit cells containing fewer than 6 cation sites were minimized with respect to the ion positions and volume (keeping cubic relationships between cell parameters.) During the minimization, some unstable configurations relax by moving an oxygen ion to a neighboring vacant site. These structures should not be included in the fitting procedure and were discarded. After elimination of the unstable structures, 188 structures remained. Forty-one ECI were fit to the 118 structures with excitation energies (energy above the ground state) less than 0.65 eV/cation. The energy cutoff increases the convergence of the cluster expansion[5]. The resulting ECI are shown in figure 1.

With the fitted ECI, a Monte Carlo simulation was carried out by starting the system at high temperature in a completely disordered arrangement at x = 0.5 and cooling in steps of 50 K. The energy vs. temperature curve is shown in figure 2.

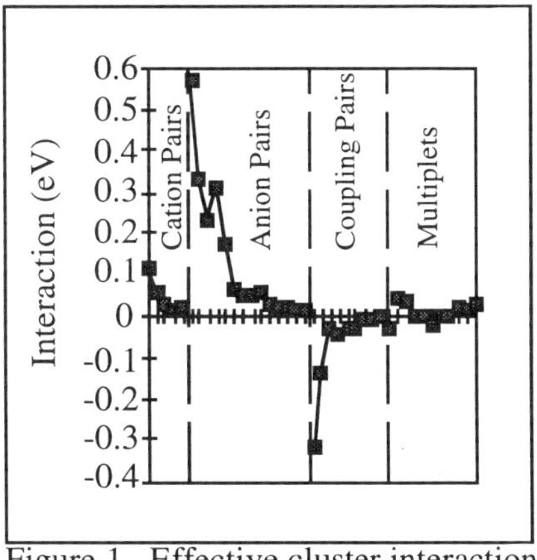

Figure 1. Effective cluster interactions fitted to energies of 118 structures computed with empirical pair potential parameters.

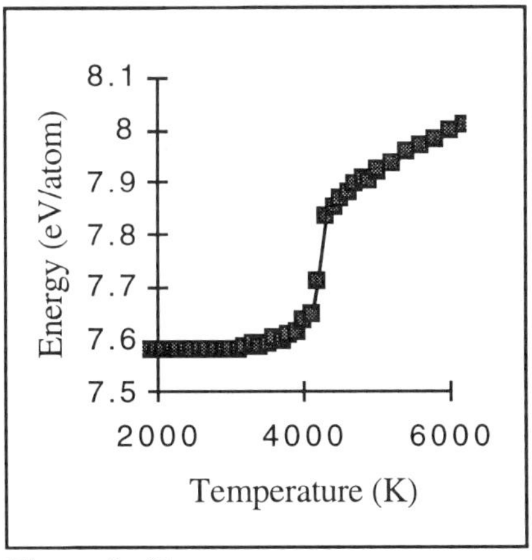

Figure 2. Energy vs temperature at x = 0.5.

The sharp jump in energy at around 4000 K is the result of a phase transition from the defect fluorite structure to pyrochlore structure. Although the transition temperature is higher than the experimental result, the fact that the model system transforms to the pyrochlore structure strongly suggests that the potential model will provide at least a reasonable qualitative description of the system. Whereas one published phase diagram[7] suggests there are two transitions (implying weak coupling), the simulation indicates only one phase transition, and strong coupling between the anion and cation sublattices. Figure 3 shows the relative probabilities of two different ion arrangements on a 5-point cluster. The probabilities are relative to those that would occur in a completely random mixture on each sublattice These plots show that the vacancies have a strong tendency to sit in Zr tetrahedra above and below the transition. However, above the transition, the vacancies also preferentially sit in tetrahedra made up of three Zr ions and one Gd ion, which are not found in the perfectly ordered pyrochlore structure.

The behavior at low concentrations of Gd_2O_3 was studied at a temperature of 2000 K. Figure 4 shows the mean squared displacement of the oxygen ions in

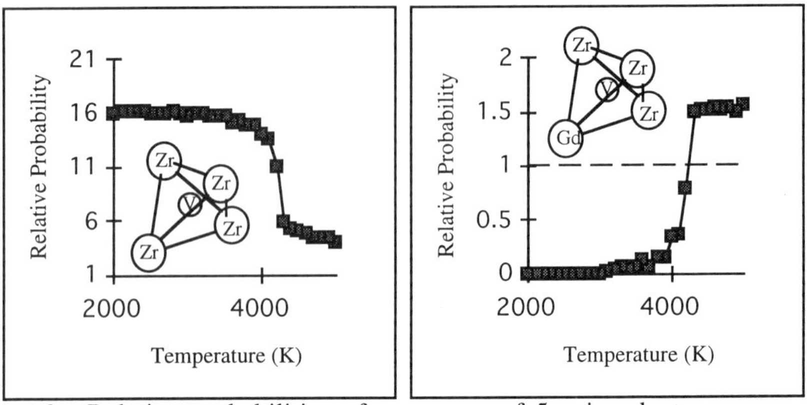

Figure 3. Relative probabilities of two types of 5-point clusters as a function of temperature at x = 0.5.

the Monte Carlo simulation as a function of the concentration for three different cation arrangements. First, with cations allowed to equilibrate (squares), there is sharp drop in the oxygen displacement at $x \approx 0.015$. Second, with the cations fixed in a completely random state (triangles), the anion displacement follows the same trend until about 0.015, then continues to increase until it levels off at $x \approx 0.04$. Finally, when the cations are first allowed to equilibrate, then fixed (circles), there is a very sharp drop in the anion displacement at $x \approx 0.015$. This indicates that the anions are strongly coupled to the cation lattice, and that the association is stronger when the cations are allowed to equilibrate. In figure 5, the relative probabilities for two representative 5-point clusters are plotted as a function of concentration for the completely equilibrated case. There is a change from a low concentration region in which the anion vacancies seem to be associated with a single Gd ion to a high concentration region where the vacancies are associated with the Zr ions. This suggests that the experimentally observed peak in conductivity is at least partially due to cation-vacancy association, and that the degree of association depends strongly on the cation configuration.

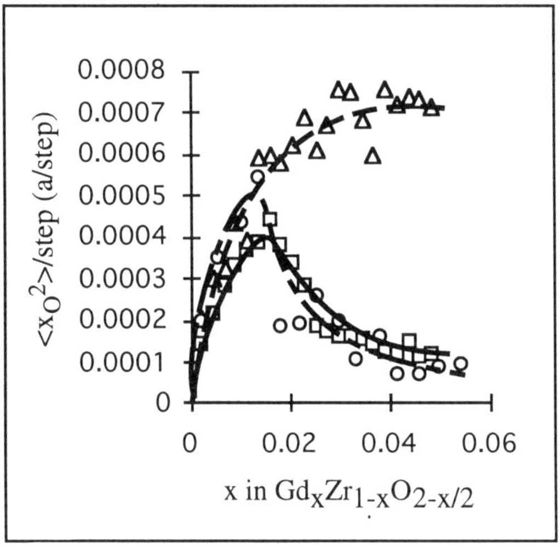

Figure 4. Mean squared oxygen displacement as a function of composition at 2000 K for three different cation conditions. Squares are for cations in equilibrium. Triangles are for cations fixed in a completely disordered arrangement. circles are for cations equilibrated, then fixed.

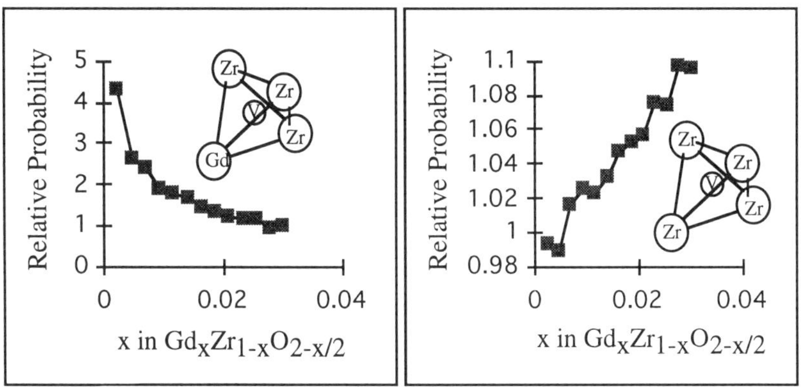

Figure 5. Relative probabilities as a function of concentration at 2000 K for two 5-point clusters in the fluorite lattice.

CONCLUSION

The $Gd_xZr_{1-x}O_{2-x/2}$ system has been modeled with a lattice model that can describe the defect behavior over nearly the entire composition range. A large number of energy minimizations using simple empirical potentials to represent the energy of the system were used as input to the model. The results indicate that there is strong coupling between the anion and cation sublattices. The strong coupling influences the behavior of the system in two ways. First, there is only one transition between the pyrochlore phase at low temperature to the defect fluorite structure at high temperature. Second, at low concentrations of Gd_2O_3, the mean squared displacement of the oxygen ions has a peak at x = 0.015 and depends strongly on the arrangement of the cations with x > 0.015. The arrangement of the vacancies with respect to the cations is strongly dependent on composition and temperature. As the system goes through the transformation from the pyrochlore to defect fluorite structure, the vacancies change from almost exclusively sitting in Zr tetrahedra in the pyrochlore phase to sitting in both Zr and three Zr - one Gd tetrahedra. Similarly, at fixed temperature, as the Gd_2O_3 concentration is increased from pure zirconia, the vacancy behavior changes from association with Gd ions to association with Zr ions. This result is in agreement with the experimental observation that at intermediate concentrations the vacancies tend to sit near Zr, rather than Gd ions[8].

REFERENCES

1. F. Ducastelle, *Order and Phase Stability in Alloys*. Cohesion and Structure, ed. F.R. de Boer and D.G. Pettifor. Vol. 3. North-Holland, 1991.
2. G. Ceder, M. Asta, and D. de Fontaine, "Computation of the Phase Diagram and Local Oxygen Configurations for $YBa_2Cu_3O_z$ with z between 6.5 and 7: The OIII Phase," *Physica C*, **177**, 106-114 (1991).
3. G. Ceder, P.D. Tepesch, A.F. Kohan, G. Garbulsky, C. Coley, H.T. Stokes, L.L. Boyer, M.J. Mehl, and B.P. Burton, "First-principles computation of the CaO-MgO phase diagram: Importance of the oxygen size effect"; This proceedings,1995.
4. P.D. Tepesch, G.D. Garbulsky, and G. Ceder, "Model for the configurational thermodynamics in ionic systems," *Phys. Rev. Lett*, **74**[12], 2272-2275 (1995).
5. G. Ceder, G.D. Garbulsky, and P.D. Tepesch, "A convergent real space cluster expansion for configurational disorder in ionic systems," *Physical Review B*, in press (1995).
6. M.A. Subramanian, G. Aravamudan, and G.V. Subba Rao, "Oxide Pyrochlores -- A Review"; p. 55-143 in *Progress in Solid State Chemistry* Vol. 15. Pergamon Press, 1983.
7. T. Moriga, S. Emura, A. Yoshiasa, S. Kikkawa, F. Kanamaru, and K. Koto, "X-ray and raman study on coordination states of fluorite- and Pyrochlore-type compounds in the system ZrO_2-Gd_2O_3," *Solid State Ionics*, **40/41**, 357-361 (1990).
8. P. Li, I.-W. Chen, and J.E. Penner-Hahn, "Effect of dopants on zirconia stabilization--an X-ray diffraction absorption study: I, trivalent dopants," *Journal of the American Ceramic Society*, **77**[1], 118-28 (1994).
9. S.B. Adler and J.W. Smith, "Effects of long-range forces on oxygen transport in yttria-doped ceria: simulation and theory," *Journal of the Electrochemical Society Faraday Transactions*, **89**[16], 3123-8 (1993).
10. A. Nakamura and J.B.J. Wagner, "Defect structure, ionic conductivity, and diffusion in yttria stabilized zirconia and related oxide electrolytes with fluorite structure," *Journal of the Electrochemical Society*, **133**[8], 1542 (1986).
11. A. Dwivedi and A.N. Cormack, "Stability of phases $CaZrO_3$ and $CaZr_4O_9$ incorporated in the system CaO-ZrO_2," *Journal of solid state chemistry*, **79**, 218-31 (1989).
12. M.P. van Dijk, A.J. Burggraaf, A.N. Cormack, and C.R.A. Catlow, "Defect structures and migration mechanisms in oxide pyrochlores," *Solid State Ionics*, **17**, 159-67 (1985).
13. R. Kikuchi, "A theory of cooperative phenomena," *Physical Review*, **81**[6], 988 (1951).

Energies and Configurations of Defects in Ga, Cr, and In-Doped CoO

S. P. Chen, M. Yan,
Theoretical Division, Los Alamos National Laboratory,
Los Alamos, NM 87545, USA

R. W. Grimes, S. Vyas
Department of Materials, Imperial College, London, UK

ABSTRACT

Atomistic simulation methods have been used to investigate the properties of point defects in CoO. The defects are substitutional ions Ga, Cr, In, and Co^{3+} holes and associated charge compensating cation vacancies. We find that while Ga, Cr and Co^{3+} ions energetically prefer to occupy second neighbor sites with respect to the vacancy, In ions sit immediately adjacent in 1st neighbor sites.

INTRODUCTION

CoO is a non-stoichiometric, metal-deficient, p-type semiconductor with the NaCl structure. It has been shown that the deficiency of Co can be as high as 1%. The dominant point defects are cation vacancies and electron holes [1-3]. Recently, the properties of point defects and defect clusters in Ga and Cr-doped CoO have been measured [4,5]. For this study, the defect is located always at Co site so we drop the position subscript of Co in the Kröger-Vink notation. The singly charged pairs, $\{Ga^{*}V''\}'$ and $\{Cr^{*}V''\}'$, were found to have binding energies of the order of 0.5 eV. The neutral triplets, $\{Ga^{*}V''Ga^{*}\}^{x}$, were found to have binding energies of the order of 1 eV. In this paper, we present the calculated results of properties of point defects and defect clusters using a shell model [6,7,8].

Calculation Methods

Detailed descriptions of the method is described elsewhere [6-9], thus only a summary is provided here. The ions interact via Coulomb interactions and short-range Buckingham potentials. The cutoff distance for the Buckingham potentials is 20 Å. In and O are treated as polarizable with shell models. The charge of the ions is split into shell charges and core charges. The shell charges interact with their cores through a spring, k. The interactions between atoms are described by the Buckingham potentials and Coulomb interactions:

$$\phi(r) = A \exp(-r/B) - C/r^6 + q_1 q_2/r$$

The parameters of the shell model were fitted so as to reproduce the crystal structure, elastic constants, low and high frequency dielectric constants of appropriate oxides. The resulting parameters are presented in Table 1. The defected structures were allowed to relax to the lowest energy states using a Newton-Raphson procedure [10].

Table 1. The fitted parameters for the shell model, where c and s stand for core and shell, respectively.

species	A (eV)	B (Å)	C ($eV/Å^6$)
Co(c)-O(s)	778.02	0.3301	-
Co^{3+}(c)-O(s)	1156.62	0.3087	-
Ga(c)-O(s)	1625.72	0.3019	-
Cr(c)-O(s)	1204.18	0.3165	-
In(s)-O(s)	1495.65	0.33272	4.3255
O(s)-O(s)	9547.96	0.21916	32.0
	k, spring constans		
In(c)-In(s)	1680.0		
O(c)-O(s)	6.3		
	Charges (q_i)		
Co(c)	2.0		
Co^{3+}(c)	3.0		
Ga(c)	3.0		
Cr(c)	3.0		
In(c)	9.1		
In(s)	-6.1		
O(c)	0.04		
O(s)	-2.04		

Results

The calculated results for the point defects and defect clusters, as shown in Fig. 1, are presented in Tables 2 to 4. Table 2 shows the isolated point defect energies, the singly charged pairs at the second nearest neighbor at <110>, the third nearest neighbor i.e. the <200> lattice site, and the fourth nearest neighbor at <211>. The energies for the singly charged pairs are measured relative to the corresponding isolated point defects. Table 2 clearly shows that the Co(3+), Cr(3+) and Ga(3+) ions prefer the third nearest neighbor position at <200>. The binding energies are 0.80 eV for Co, 0.71 eV for Cr and 0.68 eV for Ga. The corresponding experimental binding energies are 0.40 eV for Cr and 0.50 eV for Ga [4,5]. The agreement is fairly good. However, the In(3+) ion prefers the second nearest neighbor <110> site with a binding energy of 0.77 eV. Unfortunately no experimental data are available to test these proposal. However, these results strongly suggest that In ions are different from Ga , Cr and Co ions. We have recalculated the energies without allowing the In ion to be polarizable, but the result remained the same. Clearly experimental data would be particularly interesting to to confirm locating the position of In ions in CoO relative to the Co vacancies.

Table 3 shows the calculated results for neutral triplets consisting of either two Ga, Cr, Co, or In ions and a Co vacancy. The cations are located either at the second nearest neighbor or the third nearest neighbor distance. The results indicate , again, that the Ga, Cr and Co ions prefer the third neighbor sites. These ions are located symmetrically about the central Co vacancy. The binding energy for $\{Co^*V''Co^*\}^X$, $\{Cr^*V''Cr^*\}^X$, and $\{Ga^*V''Ga\}^X$ are 1.47 eV, 1.31 eV and 1.24 eV, respectively. The experimental binding energy for the $\{Ga^*V''Ga\}^X$ cluster was determined to be between 0.76 eV to 1.1 eV (for 1100 C and 1350 C) for Ga-doped CoO. While the binding energy for $\{In^*V''In^*\}^X$ is 1.40 eV. Similarly, the In ions are located symmetrically about the Co vacancy.

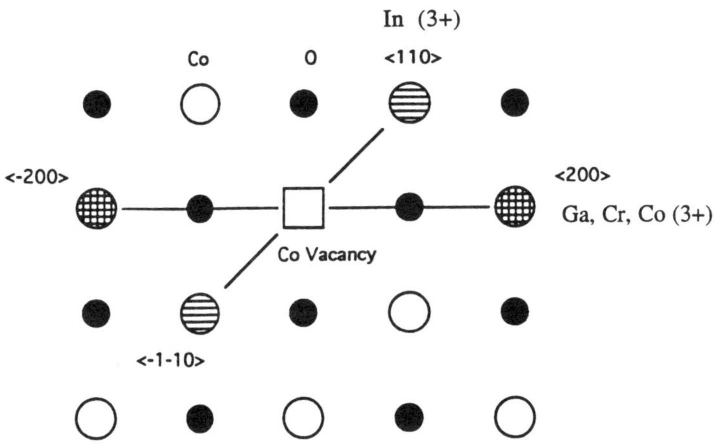

Fig. 1 The lowest energy configuration of cations relative to the Co vacancy in Ga, Cr, and In-doped CoO.

Table 2 Singlet pair energies (eV) as measured relative to the isolated defects.

Cluster	Position of metal ion with respect to vacancy			
	<110>	<200>	<211>	isolated defects
{Co*V"}'	-0.41	-0.80	-0.33	-7.23
{Cr*V"}'	-0.48	-0.71	-0.36	-4.47
		(Exp. -0.40)		
{Ga*V"}'	-0.49	-0.68	-0.37	-4.52
		(Exp. -0.50)		
{In*V"}'	-0.77	-0.47	-0.47	3.24

Table 3 Neutral triplet energies (eV) as measured relative to the isolated defects. The isolated defects energies for Co, Cr, Ga and In-triplets are -37.42, -31.90, -32.00 and -16.47 eV, respectively.

Cluster	Position of metal ion with respect to vacancy		
	<110>;<-1-10>	<110>;<-200>	<200>;<-200>
{Co*V"Co*}^{X}	-0.63	-1.04	-1.47
{Cr*V"Cr*}^{X}	-0.80	-1.05	-1.31
{Ga*V"Ga*}^{X}	-0.82	-1.03	-1.24
			(Exp. -0.76 ~ -1.1)
{In*V"In*}^{X}	-1.40	-1.10	-0.81

We also calculated the neutral triplets with mixed cations situated simultaneously around the Co vacancy. The calculated results are shown in Table 4. The results indicate that the cations seem to exhibit their own independent site preference irrespective to the position of the other cations. The Ga, Cr and Co(3+) cations prefer the third nearest neighbor distances opposite to each other. In cations prefer the second nearest neighbor distance at <110>. Therefore, an In(3+) ion, vacancy and Co(3+) ion form a bending chain. The binding energies for {Cr*V"Co*}^{X}, {Ga*V"Co*}^{X} and {In*V"Co*}^{X} are 1.39 eV, 1.36 eV and 1.42 eV, respectively. By comparing the binding energies of the neutral triplets in Table 3 and 4, we found that the mixed cations triplets exhibit almost exactly the binding energies as the sum of their equivalent homonuclear triplet components. Therefore, we predict that in Ga, Cr, and In-doped CoO the mixed neutral triplets will show no enthalpy preference over the pure triplets. For a greater discussion of cluster binding in oxides see [11].

Table 4 Mixed neutral triplet energies (eV) as measured relative to the isolated defects. The isolated defects energies for Cr, Ga and In triplets are -34.66, -34.71 and -26.95 eV, respectively.

Cluster	Position of metal ion with respect to vacancy		
	<110>;<-1-10>	<110>;<-200>	<200>;<-200>
{Cr*V"Co*}^{x}	-0.71	-1.12	-1.39
{Ga*V"Co*}^{x}	-0.73	-1.14	-1.36
{In*V"Co*}^{x}	-1.02	-1.42	-1.04

Summary

From these computations, we found that trivalent cations, Ga(3+), Cr(3+), Co(3+) and In(3+), bind with Co vacancy to form singly charged pairs with binding energies of about 0.7 to 0.8 eV. These binding energies are in reasonable agreement with the experimental measurement of about 0.5 eV. While the In ions prefer the second nearest neighbor position from a Co vacancy, the other cations prefer the third nearest neighbor sites. Two cations can also form a triplet with a Co vacancy with binding energies of about 1.2 to 1.5 eV. These values are in fair agreement with the 0.8 to 1.1 eV measured from the tracer diffusion experiments in Ga-doped CoO.

Acknowledgment: We would like to thank the support of the Division of the Materials Science, Office of the Basic Energy Sciences of U. S. Department of Energy.

References:

1. R. Dieckmann, Z. Phys. Chem., 107, 189 (1977).
2. W. K. Chen and N. L. Peterson, J. Phys. Chem. Solids, 41, 647 (1980)
3. K. Hoshino and N. L. Peterson, J. Phys. Chem. Solids, 46, 229 (1985)
4. R. Schmackpfeffer and M. Martin, Phil. Mag. A, 68, 747 (1993)
5. M. Martin, in "Defects in Insulating Materials", Proc. XII Int. Conf. 16-22 August 1992, Nordkirchen, Germany, p301-304.
6. M. Yan, S. P. Chen, T. E. Mitchell, D. H. Gay, S. Vyas, R. W. Grimes, Phil. Mag. A. (1995), in press.
7. J. D. Gale, GULP (General Utility Lattice Program), Royal Institution of Great Britain, 1993.
8. B. G. Dick and A. W. Overhauser, Phys. Rev., 112, 90 (1958).
9. J. H. Harding, Advances in Ceramics, vol. 23, "Nonstoichiometric Compounds", The American Ceramic Society, p. 229 (1987).

10. W. H. Press, S. A. Teukolsky, W. T. Vetterling, and B. P. Flannery, p. 426 in "Numerical Recipes in C", 2nd edition, Cambridge U. Press, New York (1994).
11. A. Pryde, S. Vyas, R. W. Grimes, J. Goodnew, J. Sommers, R. Wang, Phys. Rev. B. in press.

DEFECT SOLID STATE CHEMISTRY OF LANTHANUM MAGNESIUM HEXAALUMINATE

Jae-Gwan Park and A. N. Cormack
New York State College of Ceramics at Alfred University, Alfred, NY 14802

INTRODUCTION

Lanthanum magnesium hexa-aluminate (LMA, ideally $LaMgAl_{11}O_{19}$), having a structure close to that of magnetoplumbite (MP, $PbFe_{12}O_{19}$) (space group $P6_3/mmc$) [1], is currently of interest as a high-efficiency host material for the active elements of solid-state lasers [2]. The luminescent properties of substituted rare earth elements in LMA are affected by the interaction of some unknown point defects with the active elements, leading to unwanted optical absorption.

It is obvious that, due to the complexity of the basic crystal structure of MP type compounds, detailed information about their point defect properties is not readily available from experiment. In this paper, as part of our systematic studies on the structure, stoichiometry, and stability in hexa-aluminates, we investigate the defect energetics and nonstoichiometry in LMA, using theoretical, computer-based atomistic simulation techniques. The site preference of Mg in LMA and the effect of Mg distribution on the MP structure are also discussed.

COMPUTATIONAL TECHNIQUES

The lattice simulation techniques employed in this work have been described in a number of previous works for a wide range of ceramic oxides. In brief, the simulations are formulated within the framework of the Born model which treats the ionic solid as a collection of point ions with short range forces acting between them. The lattice interactions, as defined through interatomic pair potentials, are calculated explicitly and ions are relaxed to zero force using a second derivative Newton-Raphson minimization procedure. Details of the procedure have been outlined by Cormack [3]. The electronic polarization effects of the lattice are taken into account by means of the Dick-Overhauser shell model [4]. The treatment of the defective lattice follows a generalized Mott-Littleton procedure which employs the two region strategy [5]. The interatomic potentials and shell model parameters used in this study are taken from the compilations of other workers [6-8]. The viability of these potential models in the simulation of hexa-aluminates was fully discussed recently [9].

RESULTS AND DISCUSSION

Site Preference and Distribution of Mg in LMA

The structure of $LaMgAl_{11}O_{19}$ can be derived from that of $SrAl_{12}O_{19}$. Along with the substitution of a divalent Sr ion by a trivalent La ion, the charge compensation is accomplished by replacing an Al ion with an Mg ion. The question is: which Al site does the Mg ion occupy? There are five distinct crystallographic Al sites in the aluminate magnetoplumbite structure: one trigonal bipyramidal site (Al(2)), one tetrahedral site (Al(3)), and three octahedral sites (Al(1), Al(4), and Al(5)). With the x-ray data for $SrAl_{12}O_{19}$ reported by Lindop *et al.* [10] as the input structure for $LaMgAl_{11}O_{19}$, the equilibrated structures and lattice energies for different Mg ion distributions among the various Al sites are calculated. The lattice energies are listed in table I. The lower the lattice energy, the more favorable the configuration of the Mg ion in a structure. It is clearly revealed that the Mg ions prefer to occupy the terahedral (Al(3)) sites in MP structure as they do in the spinel structure. Morgan and Miles [11] pointed out that the normal site preferences of cations for tetrahedral or octahedral sites in spinels [12] are presumably paralleled in magnetoplumbite structure also, with the added complexity of competition for or avoidance of the nearly trigonal bipyramidal mirror plane sites. This result is in good agreement with that of our earlier simulation study [13], where the site preference of Mg in the MP structure was determined from the substitution energy in $SrAl_{12}O_{19}$.

Table I. The Effect of Mg Locations on the Lattice Energy of LMA per Unit Cell.

Model	Mg position	Lattice energy (eV)
LMA-I	Mg at Al(3) 4f 2, 3	-1955.17
LMA-II	Mg at Al(3) 4f 2, 4	-1954.91
LMA-III	Mg at Al(3) 4f 1, 2	-1955.45
LMA-IV	Mg at Al(1) 2a 1, 2	-1954.44
LMA-V	Mg at Al(2) 2b 1, 2	-1954.60
LMA-VI	Mg at Al(5) 4f 2, 3	-1950.36
LMA-VII	Mg at Al(5) 4f 2, 4	-1952.38
LMA-VIII	Mg at Al(5) 4f 1, 2	-1952.57
LMA-IX	Mg at Al(4) 12k 3, 12	-1952.86
LMA-X	Mg at Al(4) 12k 9, 12	-1953.45
LMA-XI	Mg at Al(4) 12k 6, 12	-1953.68

It is also revealed in table I that the distribution of Mg in different tetrahedral sites has some effect on the lattice energy (let us recall that only half of the tetrahedral sites are occupied by Mg ions in the structure and that there are three crystallographically distinct distributions of Mg among the tetrahedral sites in a unit cell): There is a slight tendency for Mg ions to be separated from each other as far as possible among the tetrahedral sites in the structure.

The Effect of Mg Distribution on the Positions of the Al Ion (Al(2)) in LMA

An outstanding controversy in the crystallography of magnetoplumbite structures is whether the cation in the nearly trigonal bipyramidal site, in this case the Al(2) cation, sits on the mirror plane (2b site) or is displaced from it along the **c** axis (4e site) becoming tetrahedrally coordinated.

In table II, the equilibrated coordinates of the Al(2) cations for different distribution of Mg in $LaMgAl_{11}O_{19}$ are listed. The ideal Al(2) coordinates in MP structure are 0 0 1/4 and 0 0 3/4. We can see that the reason for Al(2) to be displaced from the ideal position on the mirror plane may be attributed to the detailed Mg distributions in the spinel block. It is revealed that, if the two neighboring tetrahedral sites, which are coupled to each other by a mirror plane symmetry, of an interspinel layer are occupied by Mg ions or by Al ions (LMA-I), the Al(2) cation is positioned at the ideal mirror plane site, otherwise it is displaced from the mirror plane site. This structural behavior can be explained mainly by the Coulombic strains imposed on the mirror plane from the tetrahedral sites occupied by either Mg or Al ions: The strains, which are generated when the two tetrahedral sites are occupied by different ions, are relieved by the displacement of the Al(2) ions. Accordingly, the Al(2) ion displaces toward the Mg ion.

Table II. The Z Parameters of Cations in the Interspinel Layer (Ideally a Mirror Plane) for Different Mg Distribution in the Tetrahedral Sites

Mg distribution	Al(2) 1	La 1	Al(2) 2	La 2
LMA-I	0.25176	0.25176	0.75176	0.75176
LMA-II	0.26193	0.25099	0.73807	0.74901
LMA-III	0.23830	0.25000	0.73830	0.75000

This result implies that the atomic position (and thus the coordination) in the peculiar trigonal bipyramidal site in MP structures is very sensitive to vacancies or substitutional aliovalent cations, and their distribution, in the neighboring spinel blocks: The atomic position in the site is likely to be determined by the atomic structure in the neighboring spinel blocks.

Defect Energetics in LMA

Since not only the differences in lattice energies of the three distinct structures, LMA-I, LMA-II, and LMA-III, having different Mg distributions, are not very great, but also preliminary defect energy calculations reveal that some defects greatly favor a particular cation distribution in the tetrahedral sites in neighboring spinel blocks, the defect energetics and nonstoichiometry in LMA are investigated through the calculations for the formation energies of defects in each of the three equilibrated unit cell structures.

Table III. Calculated Defect Energies for the Basic Atomistic Defects. (Units in eV.)

Defect	LMA-I	LMA-II	LMA-III	Overall
La vacancy (V_{La})	37.53	38.79	38.89	**37.53**
La interstitial (La_i)	-21.72	-24.09	-24.76	**-24.76**
Al vacancy (V_{Al})	52.87	54.60	54.64	**52.87**
Al interstitial (Al_i)	-44.61	-46.04	-43.96	**-46.04**
Mg vacancy (V_{Mg})	29.47	30.70	29.17	**29.17**
Mg interstitial (Mg_i)	-16.83	-18.60	-16.28	**-18.60**
O vacancy (V_O)	22.21	21.86	23.05	**21.86**
O interstitial (O_i)	-15.15	-17.38	-13.83	**-17.38**
La Frenkel*	7.91	7.35	7.07	**6.39**
Al Frenkel*	4.13	4.28	5.34	**3.42**
Mg Frenkel*	6.32	6.05	6.45	**5.29**
O Frenkel*	3.53	2.24	4.61	**2.24**
Schottky*	2.90	3.38	4.04	**2.69**

* Energy per constituent defect

Intrinsic disorder: Within the three equilibrated crystal structures, vacancy and interstitial formation energies are calculated for each of the possible species. The energies are given in table III. The lowest formation energies for specific defects among the three structures are rewritten in the fourth column of the table, as overall defect energies in an LMA crystal in which the three kinds of Mg distributions are mixed. The Schottky and Frenkel defect formation energies per constituent defect are determined from the calculated energies for cation and anion vacancies and interstitials. It is noteworthy that because the different types of disorder involve varying numbers of defects, comparison in terms of energy per defect is essential. The thermodynamic grounds for this have been discussed elsewhere [14].

Our calculations show that oxygen Frenkel disorder is the dominant defect mode to be expected in LMA. The Schottky disorder, which was suggested by Abrahams *et al.* [15], may also be existent. We can see that the formation energies for the basic atomistic defects (especially, for the interstitial defects) are greatly influenced by the distribution of Mg. In other words, the interstitial defects strongly favor a particular distribution of Mg. For instance, the oxygen interstitial will lead to ordering of Mg (or Al) in the tetrahedral sites around it: The interstitial oxygen prefers strongly Al ions rather than Mg ions in its neighboring tetrahedral sites. Therefore, it is of great importance to consider the three structures for defect energetics in LMA.

Probable defect processes involved in nonstoichiometric LMA: There are a number of possible nonstoichiometric formulas related to $LaMgAl_{11}O_{19}$. Each formula describes different ways in which the nonstoichiometry may be accommodated. For a particular nonstoichiometric formula, several defect models (that is, charge compensation mechanisms) are also possible. Since the associated defect processes depend on the energetics of the various defect species concerned, it is worthwhile to calculate the enthalpies of the various reactions for the different formulas of nonstoichiometry to see whether a clear favorite emerged.

Table IV. Calculated Defect Energies for Substitution. (Units in eV.)

Defect	LMA-I	LMA-II	LMA-III	Overall
Al_{La}	-9.70	-12.53	-9.34	**-12.53**
Al_{Mg}	-29.28	-29.71	-29.02	**-29.71**
Mg_{Al}	28.64	28.58	29.19	**28.58**
Mg_{La}	14.75	14.31	15.22	**14.31**
O_{La}	22.54	26.68	26.46	**22.54**

Formation energies of some antisite defects, which may be involved in the defect processes, are listed in table IV. It is shown that the formation energies of the cation-cation substitutional defects (A_B) are lower than the sum of the energies of constituting vacancy-interstitial pairs (A_i + V_B). However, it is revealed that, unlike these cation-cation antisite defects, the O_{La} defect is improbable, at least, as a simple point defect in LMA, since its formation energy is greater than the sum of the energies of the individual defects V_{La} and O_i. The site potential of the La sites (as, indeed, for any cation site) usually precludes occupation of that site by a negatively charged ion, so it is not surprising that a simple O_{La} defect should be energetically unfavorable. It is noteworthy that the formation energy of Al_{La} defect is higher than the energy sum of the two defects Al_{Mg} and Mg_{La}: The simple point defect Al_{La} is expected to be also improbable. Consequently, we consider the antisite defects Al_{Mg}, Mg_{Al}, and Mg_{La} (instead of their corresponding vacancy-interstitial pairs) and the defect pair Al_{Mg} + Mg_{La} (instead of Al_{La}) in calculations of the enthalpies for the reactions where they are involved. Calculated lattice energies (per formula unit) of some oxides involved in the defect processes are: MgO = -40.44 eV; Al_2O_3 = -158.78 eV; La_2O_3 = -124.14 eV; and $LaAlO_3$ = -141.87 eV.

Typical defect processes which are found to be energetically favorable are listed in table V along with their reaction enthalpies. (A full description on the defect processes is given in one of our recent papers [16].) The mechanisms of charge compensation in the reactions are implicitly represented in the defect energy expressions given in the table.

Table V. Typical Defect Processes and Their Calculated Reaction Enthalpies. (Units in eV.)

A1. $LaMgAl_{11}O_{19} + (x/2)Al_2O_3 \rightarrow LaMg_{1-3x/2}Al_{11+x}O_{19} + (3x/2)MgO$

$E_{A1} = (3/2)E(Al_{Mg}) + (1/2)E(V_{Al}) + (3/2)E(MgO) - (1/2)E(Al_2O_3) = 0.60$

A2. $LaMgAl_{11}O_{19} + (x/2)Al_2O_3 \rightarrow LaMg_{1-x}Al_{11+x}O_{19+x/2} + xMgO$

$E_{A2} = E(Al_{Mg}) + (1/2)E(O_i) + E(MgO) - (1/2)E(Al_2O_3) = 0.55$

A3. $LaMgAl_{11}O_{19} + (x/2)Al_2O_3 \rightarrow La_{1-x/5}Mg_{1-x}Al_{11+x}O_{19+x/5} + (x/10)La_2O_3 + xMgO$

$E_{A3} = E(Al_{Mg}) + (1/5)E(O_i) + (1/5)E(V_{La}) + (1/10)E(La_2O_3) + E(MgO) - (1/2)E(Al_2O_3) = 0.86$

A4. $LaMgAl_{11}O_{19} + (x/2)Al_2O_3 \rightarrow La_{1-x/7}Mg_{1-8x/7}Al_{11+x}O_{19+x/7} + (x/14)La_2O_3 + (8x/7)MgO$

$E_{A4} = (1/7)E(V_{La}) + (1/7)E(O_i) + (8/7)E(Al_{Mg}) + (1/7)E(V_{Al}) + (1/14)E(La_2O_3) + (8/7)E(MgO) - (1/2)E(Al_2O_3) = 0.78$

A5. $LaMgAl_{11}O_{19} + (x/2)Al_2O_3 \rightarrow La_{1-x/5}Mg_{1-6x/5}Al_{11+x}O_{19} + (x/10)La_2O_3 + (6x/5)MgO$

$E_{A5} = (1/5)E(V_{La}) + (6/5)E(Al_{Mg}) + (1/5)E(V_{Al}) + (1/10)E(La_2O_3) + (6/5)E(MgO) - (1/2)E(Al_2O_3) = 0.88$

B1. $LaMgAl_{11}O_{19} + xMgO \rightarrow LaMg_{1+x}Al_{11-x}O_{19-x/2} + (x/2)Al_2O_3$

$E_{B1} = E(Mg_{Al}) + (1/2)E(V_O) + (1/2)E(Al_2O_3) - E(MgO) = 0.56$

B2. $LaMgAl_{11}O_{19} + xMgO \rightarrow LaMg_{1+x}Al_{11-2x/3}O_{19} + (x/3)Al_2O_3$

$E_{B2} = (2/3)E(Mg_{Al}) + (1/3)E(Mg_i) + (1/3)E(Al_2O_3) - E(MgO) = 0.37$

The defect processes group A in table V concern the nonstoichiometries possible when Al ions are incorporated in excess into the ideal LMA structure. All the listed Al-excess defect processes are likely to be existent, from the viewpoint of reaction enthalpies, even though the degrees of nonstoichiometry at equilibrium will be different and be most significant in the two processes A1 and A2. Therefore, the reaction path for the Al-excess processes is expected to be various and will be determined by kinetics or by experimental conditions, such as starting composition, oxygen partial pressure, and so forth. It is important to note that all these promising Al-excess defect processes involve the Al_{Mg} defect: It is evident in the reaction enthalpies that this defect plays a key role in the Al excess nonstoichiometric processes in LMA. The compensating defect species for the Al_{Mg} are expected to be O_i, V_{Al}, V_{La}, or the combinations of them. Among these promising Al-excess defect processes, the processes A2, A3, and A4 can also be regarded as oxygen-excess processes which are likely to occur. It is noteworthy that the excess oxygen in these processes prefers the O_i (rather than O_{La}) which has lowest formation energy when it is positioned at a center of the spinel block without containing Mg ions.

The defect processes group B in table V concern the nonstoichiometries which are possible when Mg ions are incorporated in excess into the ideal LMA structure. The two listed Mg-excess defect processes are found to be energetically favorable. It is noteworthy that the defect Mg_{Al} plays an important part in the Mg-excess defect processes, as the Al_{Mg} does in the Al-excess processes. The compensating defect for the Mg_{Al} is expected to be Mg_i or V_O, which may be determined by experimental conditions.

CONCLUSIONS

Computer-based atomistic simulation methods are applied to address quantitatively the defect energetics and crystal chemistry of LMA. The tetrahedral site preference of Mg in the magnetoplumbite structure is determined by calculating lattice energies for different Mg ion distributions in LMA. The position of trigonal bipyramidal Al(2) cations can be displaced from the ideal mirror plane site along the **c** axis as well as can be positioned on the ideal site, depending on the detailed Mg distributions among the tetrahedral sites in the neighboring spinel blocks. It is revealed that the intrinsic and extrinsic disorders in LMA are much influenced by the distribution of Mg among the tetrahedral sites in the structure. Several feasible defect processes in non-stoichiometric LMA are determined from the enthalpies of the quasi-chemical reactions for the processes with simple defect energies. It is revealed that the defects Al_{Mg} and Mg_{Al} play a key role in the Al-excess and Mg-excess processeses, respectively.

ACKNOWLEDGMENTS

The U. S. Department of Energy, Office of Basic Energy Sciences is thanked for financial support under grant number DE-FG02-91ER45451. Some of the calculations reported here were performed at the Cornell Theory Center, which is funded in part by New York State, the National Science Foundation and IBM corporation.

REFERENCES

1. A. Kahn, A. M. Lejus, M. Madsac, J. Thery, D. Vivien, and J. C. Bernier, "Preparation, Structure, Optical, and Magnetic Properties of Lanthanide Aluminate Single Crystals ($LnMAl_{11}O_{19}$)," *J. Appl. Phys.*, **52**, 6864-69 (1981).
2. L. D. Schearer, M. Leduc, D. Vivien, A. M. Lejus, and J. Thery, "LNA: A New Neodymium Laser Tunable around 1.05 and 1.08 μm," *IEEE J. Quant. Electron.*, **QE22**, 713-17 (1986); B. Viana, G. Aka, D. Vivien, A. M. Lejus, J. Thery, A. Derory, J. C. Bernier, C. Garapon, and G. Boulon, "Absorption, Fluorescence, and Electron Spin Resonance Investigation of Tervalent Cerium Activated $LaMgAl_{11}O_{19}$," *J. Appl. Phys.*, **64**, 1398-1404 (1988).
3. A. N. Cormack, "A Perfect Lattice Approach to Nonstoichiometry," *Solid State Ionics*, **8**, 187-92 (1983).
4. B. G. Dick and A. W. Overhauser, "Theory of the Dielectric Constants of Alkali Halide Crystal," *Phys. Rev.*, **112**, 90-103 (1958).
5. C. R. A. Catlow and W. C. Mackrodt, "Theory of simulation methods for lattice and defect energy calculations in crystals," in Computer Simulation of Solids edited by C. R. A. Catlow and W. C. Mackrodt, pp. 3-20, Springer-Verlag, Berlin (1982).
6. G. V. Lewis and C. R. A. Catlow, "Potential Models for Ionic Oxides," *J. Phys. C: Solid State Phys.*, **18**, 1149-61 (1985).
7. V. Butler, C. R. A. Catlow, B. E. F. Fender, and J. H. Harding, "Dopant Ion Radius and Ionic Conductivity in Cerium Dioxide," *Solid State Ionics*, **8**, 109-113 (1983).
8. C. R. A. Catlow, "Point Defect and Electronic Properties of Uranium Dioxide," *Proc. R. Soc.*, A, **353,** 533-61 (1977).
9. J. G. Park and A. N. Cormack, "Potential Models for Multi-Component Oxides: Hexa-Aluminates," *Phil. Mag.* B, in press.
10. A. J. Lindop, C. Matthews, and D. W. Goodwin, "The Refined Structure of $SrO{\cdot}6Al_2O_3$," *Acta Cryst., B* **31**, 2940-41 (1975).
11. P. E. D. Morgan and J. A. Miles, "Magnetoplumbite-Type Compounds: Further Discussion," *J. Am. Ceram. Soc.*, **69** [7] C157-C159 (1986).
12. H. St. C. O'Neil and A. Navrotsky, "Simple Spinels: Crystallographic Parameters, Cation Radii, Lattice Energies, and Cation Distribution," *Am. Mineral.*, **68**, 181-94 (1983).
13. L. Xie and A. N. Cormack, "Cation Distribution in Magnetoplumbite and β"-alumina Structures," *Materials Letters*, **9**, 474-79 (1990).
14. A. N. Cormack, "Defect Processes in Ceramics," *Adv. Solid-State Chem.*, **3**, 63-98 (1993).
15. S. C. Abrahams, P. Marsh, and C. D. Brandle, "Laser and Phosphor Host $La_{1-x}MgAl_{11+x}O_{19}$ (x = 0.050): Crystal Structure at 295 K," *J. Chem. Phys.*, **86**, 4221-27 (1987).
16. J. G. Park and A. N. Cormack, "Defect Energetics and Nonstoichiometry in Lanthanum Magnesium Hexa-aluminate," in manuscript.

Glass Processing

OUTLOOKS FOR MATHEMATICAL MODELLING OF THE GLASS MELTING PROCESS

H. de Waal, TNO Institute of Applied Physics, Delft, The Netherlands

1. INTRODUCTION

Mathematical modelling is nowadays a standard tool for major producers of float glass, T.V. glass and fiberglass. Also for container glass furnaces, glass tank modelling proves to be a valuable method to optimize process conditions. Mathematical modelling is no longer just a way to visualize the flow patterns and to provide data on heat transfer. It can also predict glass quality in relation to process parameters, because all chemical and physical phenomena are included in the latest generation of models, based on experimental and theoretical research on these phenomena [1-6].
However, to provide accurate results, a mathematical model needs realistic input parameters and boundary conditions, such as:

- rheology and heat transfer in molten glass and batch blanket;
- heat input by radiation from the combustion chamber;
- furnace shape and dimensions, including changes due to corrosion
- details on bubbling, boosting and flow barriers.

Often it is found that our knowledge of these parameters is insufficient. If we can overcome this barrier, the way to still other applications of mathematical modelling is open, so that it can become a tool for furnace operators and "trouble shooters". We will try to elucidate this theme further in this paper.

2. THE USE OF MATHEMATICAL GLASS TANK MODELLING

About 15 years ago mathematical modelling was merely used in order to obtain a more fundamental understanding of the flow conditions and mixing behaviour in glass melting tanks, effected by bubbling and temperature gradients.
Since the late eighties glass companies have started to use detailed models to optimize the process in order to increase pull rates and/or improve glass quality. The most suitable firing conditions, the boosting and bubbling parameters, the glass depth and the batch composition can be derived to obtain an optimum glass quality. The output data of such mathematical modelling studies are for example:

- temperature distribution in three dimensions;

- velocity vectors at every position in the melt;
- length, thickness and shape of batch blanket;
- local heat transfer rates and energy balances;
- residence time distribution;
- glass quality numbers on melting, fining and homogenisation.

Applications of the model are, for example, the prediction of the effect of (a change in) glass colour, which determines the radiative heat transmission in the melt, on the temperature distribution in the melt. The effect of bubbling or electrical boosting to correct these temperature levels for coloured glass can be established quite accurately by today's models.

At this moment 3-D modelling also becomes an important tool for the design of conventional glass furnaces. Combination of glass tank and combustion chamber modelling [7,8,9] is an important tool for the design of new systems like oxy-fuel fired glass furnaces.
The models are used to find the optimum position of the burners, the number and position of electrodes for electrical boosting and the optimum dimensions of the combustion chamber and melting tank.

3. VALIDATION

The rapid improvements in computer facilities have made it possible to run complicated and very detailed models in a reasonable time using a relatively fine grid cell size.
However, it cannot be overemphasized that these calculations have to be validated experimentally. Not only in the laboratory, but in real furnaces.
The results of mathematical modelling can be validated by measurements of temperatures in the glass melt using thermocouples for permanent or temporary measurements. At this moment the models are able to predict local glass melt temperatures within 10-15 °C accuracy.
Residence time distribution or minimum residence times can be measured by the use of tracers in the batch and compared with calculated values.
In addition, surface flows can be determined by the use of floaters and heat transfer can be measured by heat flux metering.
At this very moment, the prediction capability of the models with respect to the essential glass quality factors and redox state of the glass are being validated by performing redox calculations, fining and melting kinetics calculations and comparing those with statistical data on glass defects occurring in glass products and on-line measurements of the redox state in feeders or forehearths.

It is certainly not sufficient to validate the model as such. In fact, as long as we have no absolutely reliable data on glass properties and since usually our knowledge on the boundary conditions of an existing furnace is incomplete or unreliable, so-called fine-tuning is necessary in order to obtain a satisfactory match between model and reality. In principle this has to be done for each new combination of glass and furnace. After fine-tuning, the model is ready to perform parameter studies for this particular furnace, in order to find the optimum process conditions and to improve the design data on furnace dimensions, position of electrodes, flow barriers, bubblers, batch charging etc.

4. SUB-MODELS AND GLASS QUALITY

Sub-models have been developed to investigate, by mathematical modelling, the melting kinetics of the raw materials for a given furnace and given process conditions. The dissolution kinetics of sand grains [10,11] can be derived depending on the initial position in the batch and the trajectory in the melt for a given initial grain size.
Other sub-models describe the concentration distribution of gases in the melt and the redox distribution. With these models the behaviour of gas bubbles (size and composition as a function of position) can be studied [12-14]. This asks for reliable data on gas solubilities, gas diffusivities and redox reaction or fining reaction equilibria in molten glasses.

One of the most powerful techniques in nowadays modelling is the so-called "particle-tracing". With this technique it is possible to follow particles (grains, bubbles, inhomogeneities) and to study their interaction with their immediate surrounding during their travel through the furnace.
Depending on the exact position in the batch blanket, the batch particles can follow different trajectories in the tank during the melting process. By following-let us say-10.000 batch particles, starting at the charger and ending in the throat or feeder we can obtain a statistical image of the dissolution kinetics of-for instance- sand grains. Of course the same can be done with bubbles and when this is combined with the redox-distribution sub-model, we can really predict the fining behaviour of the melt.
The effect of bubbling, electrical boosting, initial sand grain size, pull rate etc. on the melting behaviour can also be established in this way.

5. RECENT DEVELOPMENTS

Glass melting is not a static process. Pull rate changes, colour and composition changes require new furnace settings.
It has recently become possible to perform model calculations for a furnace in non - steady-state conditions. This is an important step in the practical applicability of mathematical models. After all, the temperature, flow conditions and chemistry not only depends on the position in the tank but also on real time.
The dynamics of the glass melting process or the time dependency of the local temperatures, flows or melting indices can be calculated during sudden changes in the process conditions.
For instance, the time dependent change of the temperature distribution in the glass melt during pull rate changes or changes in batch composition and colour can be predicted
Simulation of time transient behaviour can be performed in order to find the optimum process parameter adjustments to reduce the time necessary for a complete

colour change so that production losses can be minimized and harmful temperature changes can be avoided.
Time transient modelling provides the possibility to predict the effect of a change in the process conditions on temperatures in the furnace, the glass quality and flow conditions. Therefore it could be used as a fundamental part in furnace control strategies.

As an illustration, figure 1 shows the effect on temperature in the throat after a change from flint to green glass with and without changing the firing conditions. In this way the optimum new process parameter settings can be found.
Another example is the transition from funnel glass to screen glass in a (fictive) TV furnace, shown in figure 2. Screen glass has a lower density, a lower thermal conductivity and a higher viscosity than funnel glass. The transition is calculated over a period of 60 hours, twice the average residence time of this furnace.

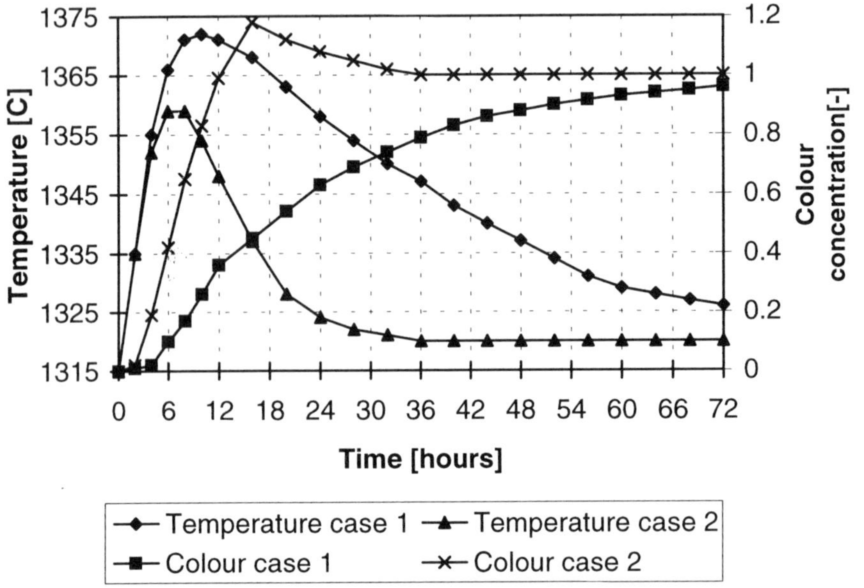

Figure 1. Effects of a colour change in a container tank from flint to green glass and increased crown temperature (+ 50 °C), on the time dependent temperature in the throat and time dependent colour of the glass (0 = pure flint, 1 = pure green).
At time t=0, the batch composition has been changed by adding the colourizing agent. In case 1, the concentration of the added colourizing agent in the batch is normal for green glass. In case 2, the concentration of colourizing agent has been doubled for the first 12 hours, in order to speed up the colouring of the flint glass still present in the tank.

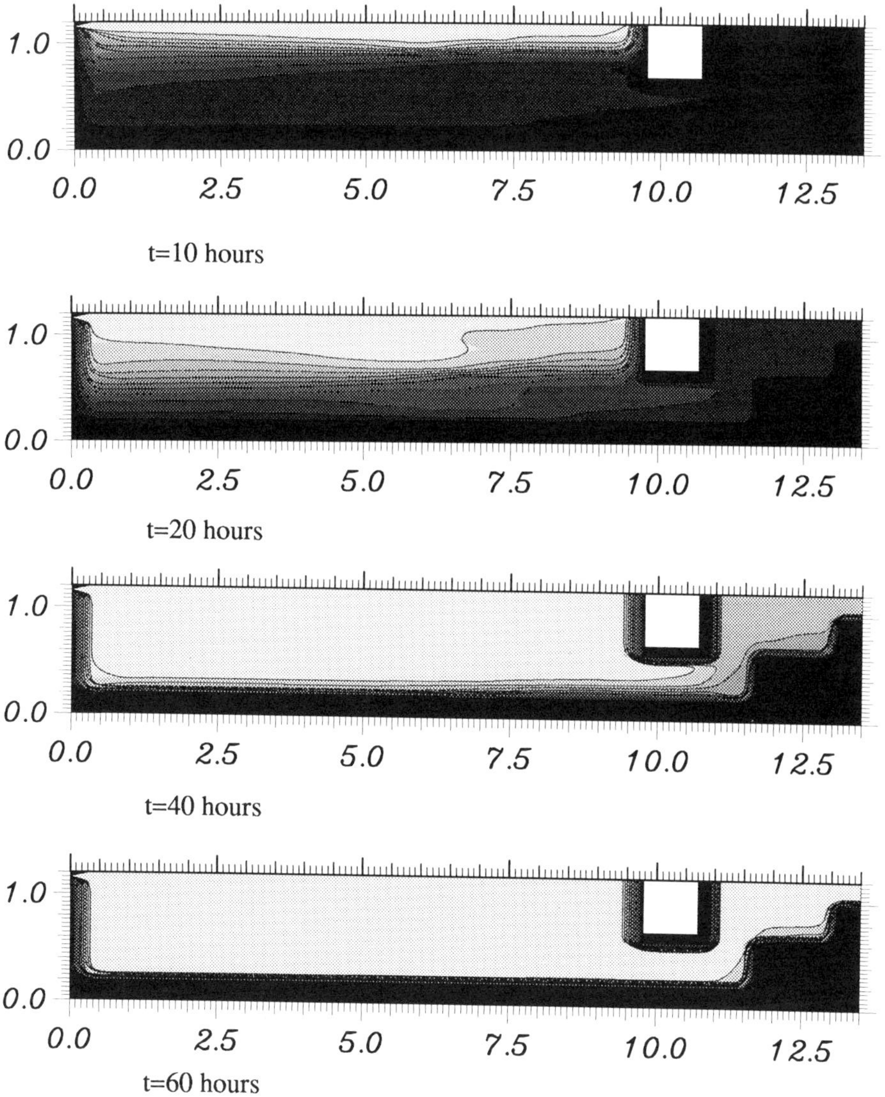

Figure 2 Transition from funnel glass to screen glass in a fictive TV glass furnace, producing 80 tons/day. The concentration distribution is shown 10, 20, 40 and 60 hours after first charging the screen glass batch composition (shown in grey).

6. FURTHER IMPROVEMENTS

Further improvements of the reliability of the present models are necessary and possible by comparing measured values of - for instance - temperatures and glass quality as a function of the position in the tank with mathematical results. The present state of the art is a model, able to predict local temperatures within 10-15 °C accuracy for most furnaces.
Often this will be sufficient, but in some cases a higher accuracy would be desirable. Availability of precise data on glass properties like the temperature - and wavelength dependent radiant heat transfer coefficient in the molten glass, volumetric thermal expansion coefficient and viscosity for certain glass compositions are a necessary condition for obtaining more accurate results.
The determination of the properties of gases in glass melts, redox reaction equilibria and dissolution kinetics of raw materials in glass melts is a target for future experimental investigations in order to improve the reliability of the sub-models for fining and melting kinetics.
In addition, new sub-models have to be developed to describe the homogenisation process, the volatilisation from the melt surface, foam formation and stability and the interaction molten glass - refractory.

7. FUTURE CHALLENGES

Potentially mathematical modelling has more to offer to glass technologists than what has been realized up to now. Especially if it is combined with other techniques, which are now being used or in development independently.
Two examples will be given here.

1. From physical and chemical analyses of defects in glass products (stones, cords, bubbles, etc.) it is sometimes possible to identify the (probable) origin of the defect and to relate it to process parameter settings or to certain sections/locations in the furnace. Empirical databases have been developed, including practical know-how, to make this technique more operational. Combination with the particle tracing technique in mathematical modelling might considerably enhance the accuracy and reliability of these identification procedures.
   It is, for instance, possible to create a statistical data-base from computer simulation, which corrrelates all kinds of defects, originating from different locations in the furnace with defects in the product or in the furnace exit.
   As an example, fig. 3 and 4 show the behaviour of oxygen bubbles, originating at the refractory-melt interface just before the throat in a sulphate-fined container tank. In fig. 3 the percentage dissolved, escaped and remaining bubbles in the forehearth is given in dependance of the original bubble size. As might be expected, large bubbles mainly escape and small predominantly dissolve or remain in the melt. Fig. 4 gives the size distribution and composition of seeds, to be found in the forehearth originating from oxygen bubbles from the same location with an initial diameter of 0.8 mm. It is interesting to see that almost all oxygen has

diffused out of the bubbles and that the remaining gas composition depends on the size of the seed. Of course this is just an example; the results very much depend on furnace conditions like redox and temperature.
In fig. 5 another example is shown: here the behaviour of air bubbles originating from the batch in a sulphate-fined furnace has been calculated. Also in this case, large bubbles mainly escape during their travel through the furnace. Small bubbles mainly dissolve. In this particular case some bubbles with an original diameter of 0.3-0.4 mm will be found in the forehearth, of course with different composition and size.

2. In future mathematical modelling will become a part of furnace control systems, in any case off-line and if computer developments continue to be as rapid as they are now, also on-line.
Time-transient calculations will predict the effect of changes in pull, batch composition, etc. on glass quality and process parameters like temperature distribution, convection flows,redox state, etc. These predictions will give valuable information to find optimum control conditions of a glass furnace. New process control concepts, like "fuzzy control" are in development at this moment, usually based on "black-box approach" or concepts, based on statistical information. In the combination of these approaches with process modelling we may very well find our real challenge for the future.

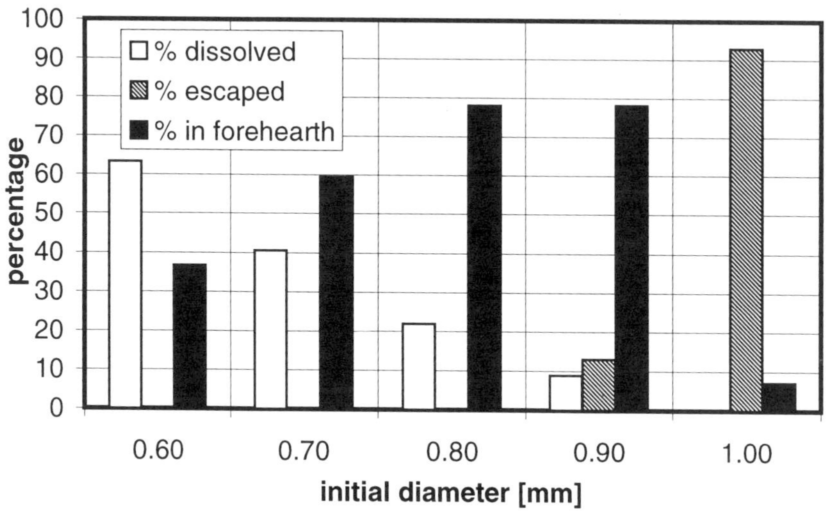

Figure 3 Percentage of bubbles that dissolve, escape or end up in the forehearth of a sulphate-fined container tank. The bubbles originally contained only oxygen and originated at the interface bottom refractory-melt just before the throat with initial diameters ranging from 0.6-1.0 mm.

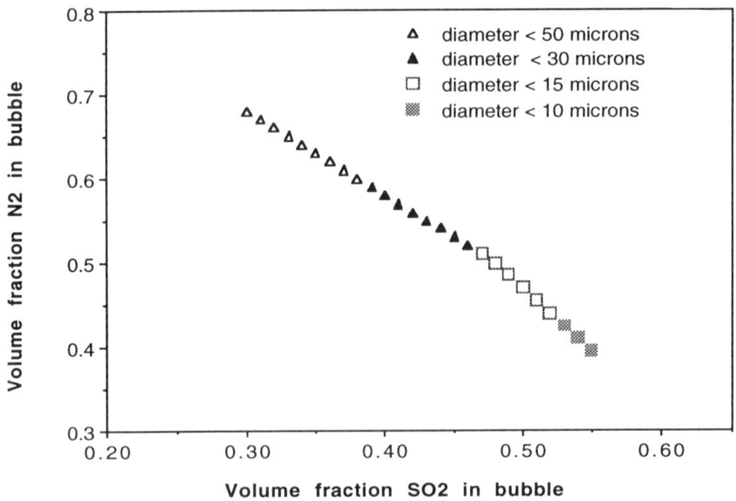

Figure 4 Calculated composition and size of seeds in a glass product, originating from 0.8 mm oxygen bubbles, generated at the refractory-glass interface at the bottom of the tank, just before the throat.

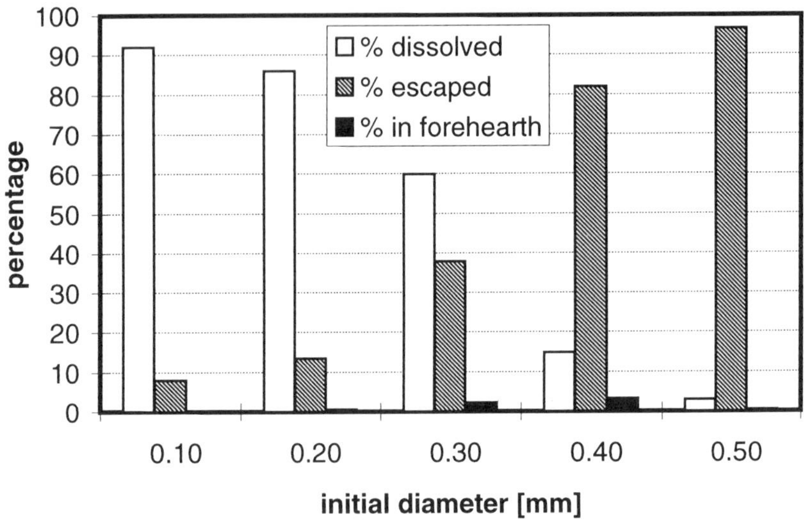

Figure 5 Percentage of air bubbles that dissolve, escape or end up in the forehearth of a sulphate-fined container tank, originating from the batch with initial diameters ranging from 0.1 to 0.5 mm.

8. CONCLUDING REMARKS

Three dimensional glass tank modelling in conjunction with sub-models describing melting kinetics or behaviour of gases and gas bubbles in a melt is becoming an important tool for the optimization of furnace design and the melting process. The influence of batch composition, furnace geometry, dimensions, forced bubbling, the position of bubblers and flow barriers, the distribution of the energy over the different burners, the action of electric boosting on the quality of the melting process or degassing process can be studied using models based on detailed knowledge of the chemical and physical processes in the glass melting tank. Extensions of the models presented here, will fulfil the need to obtain complete understanding of the influence of adjustable process parameters (like pull rate, bubbling, boosting, firing) on the presence of stones, knots and bubbles in the product, but also on homogeneity and colour.

By using particle tracing techniques, the origin of glass defects, like bubbles with a certain size and composition, or the source of unmolten material found in products, may be discovered. This could become an important tool for furnace operators and trouble shooters to identify the causes for process calamities and to localise defect sources.

Glass tank modelling in combination with glass quality sub-modelling is still in a stage of development. Further verification of the obtained results and accurate determination of relevant molten glass properties are essential to increase reliability and applicability.The combination of glass tank modelling with furnace control concepts and glass defect analysis procedures are future challenges for R&D in glass technology.

LITERATURE

1. Ungan A.; "Three dimensional numerical calculation of fining process in glass melting furnaces" 2nd Int. Conf. on Advances in the Fusion and Pro - cessing of Glass 1990, Düsseldorf, Glastechn. Ber. Sonderb.LXIIIK pp.19- 28
2. Ungan A., Viskanta R.; "Three dimensional numerical modelling of circulation and heat transfer in a glass melting process parts 1 and 2", Glastech. Ber. 60 (1987) no.3 pp. 71-78 and no. 4 pp. 115-124 .
3. Schill P. "Calculation of 3-dimensional Glassmelt Flow in Large Furnaces via Twogrid Method" 2nd Int. Conf. on Advances in the Fusion and Processing of Glass 1990, Düsseldorf , Glastechn. Ber. Sonderb. LXIIIK pp.39-47
4. de Waal H. "Mathematical modeling of the glass melting process" 2nd Int. Conf. on Advances in the Fusion and Processing of Glass 1990, Düsseldorf, Glastechn. Ber. Sonderband LXIIIK pp. 1-18
5. Simonis F., De Waal H., Beerkens R.; "Influence of furnace design and operation parameters on the residence time distribution of glass tanks, predicted by computer simulation" Collected Papers XIV Int. Congress on Glass (1986) New Dehli vol. III pp. 118-127
6. Patankar S.V.; "Numerical heat transfer and fluid flow" Hemisphere

Publishing Co. McGraw-Hill 1980

7. Hoogendoorn C.J., Post L., Wieringa J.A.; "Modeling Combustion and heat transfer in Glass Furnaces" Glastech. Ber. 63 (1990) nr. 1 pp. 7 - 128.
8. Carvalho M.G., Lockwood F.C. "Thermal Comparison of Glass Furnace Operation with Oil and Natural Gas" Glastech. Ber. 63 (1990) no. 9 pp. 233-243
9. Muijsenberg H.PH., Simonis F., Beerkens R.G.C.; "Modeling the Combustion Chamber of a Glass Furnace" Collected Papers of the "International Seminar on Mathematical Simulation in Glass Melting", Valsské Mezirrici in Czechoslovakia, 19-21 May 1991, pp. 74-83
10. Choudhary J.K.; "Free convection effects of the dissolution of a spherical particle" Proceedings of the 1th Int. Conf. on Advances in the fusion of Glass,1988 Alfred New York pp. 11.1-11.23
11. Beerkens R. "Kinetik der Quarzsandauflösung in Gemengeteppich und in Glasschmelzen" 63. Glastech. Tagung der DGG Stuttgart 1989 pp. 65-70
12. Simonis F. "Estimation of the local redox distribution in the melt by numerial flow modelling" 2nd Int. Conf. on Advances in the Fusion and Processing of Glass 1990, Düsseldorf, Glastechn. Ber. Sonderb. LXIIIK pp. 29-38.
13. Beerkens R., "Chemical Equilibrium Reactions as Driving Forces For Growth of Gas Bubbles During Refining" 2nd Int. Conf. on Advances in the Fusion and Processing of Glass 1990, Düsseldorf, Glastechn. Ber. Sonderb. LXIIIK pp. 222-242
14. Nêmec L.; "The Behaviour of Bubbles in Glassmelts part 1: Bubble size controlled by Diffusion and part 2 : Bubble size controlled by Diffusion and Chemical Reaction" Glass Technol. vol. 21 (1980) nr. 3 pp. 134 - 144

RECENT ADVANCES IN THEORETICAL MODELING OF DEPOSITION, DRYING, AND SHRINKAGE IN SOL-GEL COATING PROCESSES*

R. A. Cairncross, K. S. Chen, P. R. Schunk, C. J. Brinker, and A. J. Hurd
Sandia National Laboratories, Albuquerque, NM 87185-0827

ABSTRACT

The processing conditions during coating and subsequent drying of sol-gel films strongly affect the final film properties and microstructure. This paper is an overview of recent advances in theoretical modeling of microstructure evolution in sol-gel thin films. Computational methods in fluid mechanics, mass transfer, and solid mechanics provide insight into the relationships between processing conditions and thin-film coating properties.

1. INTRODUCTION

Sol-gel processing is a method for producing ceramics by solution chemistry. A *sol* is a liquid solution of reactive polymers or suspension of meta-stable colloidal particles. A *gel* is an elastic solid which forms from chemical reactions or colloidal aggregation of the ceramic precursors. When this *sol-to-gel* transformation occurs, the solid skeleton of the gel is filled with solvents, which must be removed by evaporation or drying. Subsequent shrinkage and capillary pressure during the drying process causes large stresses that can lead to cracking. Thus, sol-gel processing is especially suited for producing films where these stresses are minimized [1].

Most sol-gel coatings are prepared by dip-coating or spin-coating. Both of these coating processes are convenient laboratory techniques for producing thin, uniform coatings on a wide variety of substrate shapes and types. This paper focuses on the dip-coating process and the application of theoretical modeling to the various stages of this process: deposition and entrainment, drying and solidification, pore structure evolution, and stress development. Brinker et al. (1992)[2] have shown that the final properties of the ceramic coating are sensitive to the processing conditions in all of these stages. Thus, a predictive model of how processing conditions affect the microstructural evolution during the dip-coating process would enable producing films with controlled properties, such as controlled poros-

*. This research was supported in part by the Basic Energy Sciences program of the DOE and was performed under the contract DE-AC04-76-DP00789.

ity for use in ceramic membranes. This paper discusses predictions from each of these stages as an independent process. Due to brevity of the paper, details of the equations and their numerical solution methods are left to the references.

2. PREDICTIONS OF DIP COATING AND DRYING

Coating films of advanced sol-gel (ceramic) materials require precise drying control and hence complete knowledge of the details of the heat, mass, and momentum transport fields in both the coating liquid and drying gas phases. Dip coating of pure fluids is governed primarily by the balance between viscous, gravitational, and capillary forces. However in dip-coating and drying of multi-component solutions additional effects of surface tension gradients and gas phase transport are needed for accurate predictions. The differential evaporation rates of solvents, such as ethanol and water, create concentration gradients in the entrained coating solution; this causes surface tension gradients that can drive additional flow toward the drying line (as in the classical 'tears of strong wine' [3]) or away from the drying line (as in the Marangoni drying process [4]). Experimentally observed increasing evaporation rates of solvents near the drying line (due to reduced gas-phase resistance [5]) invalidates using a mass transfer coefficient to describe the gas-phase transport.

Vaporization or condensation phenomena complicate conventional analysis of free surface flows: the most simplified but realistic theory must include convection-diffusion together with the Navier-Stokes system and all boundary conditions common to free surface flows. The theoretical model developed by Schunk and Rao (1994)[6] accounts for interphase transfer of multicomponent solvents, differential evaporation and condensation, and surface tension gradients, thereby enabling the realistic analysis of drying in dip-coating processes. Most theoretical models of drying available today use lumped mass transfer coefficient theory for the gas phase transport.

3. PREDICTIONS OF DRYING AND GELATION

The short time of the deposition process (~ 10 s) establishes an important time scale for the dip-coating process and distinguishes sol-gel thin film formation from bulk (monolithic) gel formation. Often the time required to prepare a monolithic bulk gel is on the order of days to months. During dip-coating the drying stage completely overlaps the gelation and aging stages. The depositing sol (which initially is often less concentrated than sols intended for bulk gels) remains rather dilute until the final stage of the deposition process at which point it is rapidly concentrated.

Using drying and reaction theory, Cairncross et al. (1995) [7] showed that changing the drying rate can change when gelation occurs: 1) slow drying (as in bulk

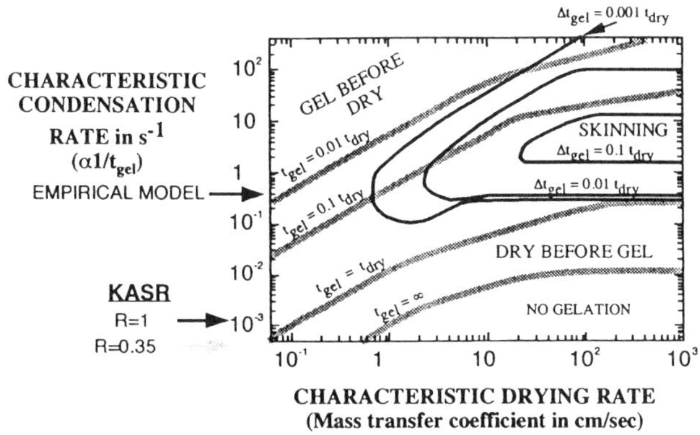

Figure 1) Predictions of drying and reaction from Cairncross et al. 1994. Varying the reaction rate and drying rate produces several types of gelation phenomena. Arrows indicate typical reaction rates using the kinetic scheme of Kay and Assink (KASR) and an empirical reaction scheme.

gels) results in gelation before solvent removal, 2) rapid drying in thick films causes a gelled film to form on the surface of the coating, 3) rapid drying in thin films causes gelation to occur after drying is completed, or 4) rapid drying in thin films can create dry un-gelled films. Figure (1) demonstrates this effect for drying silicate films [7] using the kinetic scheme of Kay and Assink (1988)[8] to predict the reaction rates. However, the reaction rates have been shown experimentally to be much more concentration dependent than those implied by Kay and Assink's theory. During the rapid concentration of reactants, the condensation reaction rate rises sharply; so the characteristic condensation rate in a rapidly drying coating is difficult to determine.

4. PREDICTIONS OF PORE-SIZE ELVOLUTION IN DRYING FILMS

In the final stage of the deposition process when a gel has formed, the capillary pressure C_p, created by tiny liquid-vapor menisci as they recede into the gel, is maximized [2]. The curvature of the menisci causes the liquid to be in tension (i.e. subambient pressure in the pore liquid) and the network to be in compression. It is the balance between the tensile capillary pressure in the liquid and the compressive elastic stress in the solid which establishes the final density and pore size of the film [9].

In a simplified analysis of drying, stress development, and pore shrinkage in a thin sol-gel film, Chen and Schunk (1994)[10] have computed predictions of the dynamic evolutions of solid-phase stress, liquid-phase pressure, and mean pore size for a model sol-gel system. Their representative results are displayed in Figure (2). The magnitude of the capillary pressure is estimated by the Young-Laplace equation:

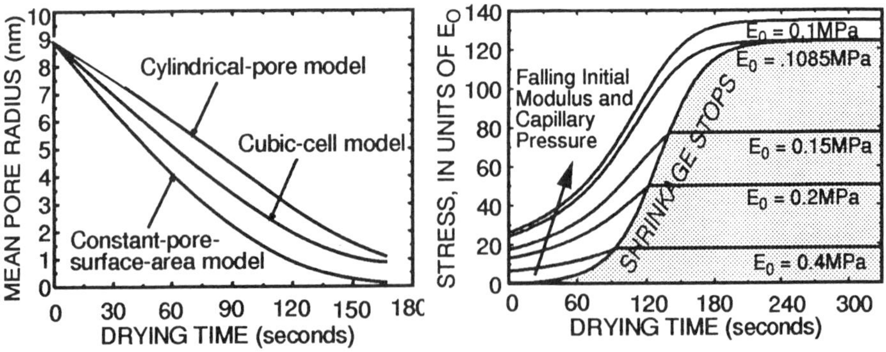

Figure 2) Predictions of pore evolution in a shrinking coating. The stress in the solid network rises as the gel shrinks, and the maximum capillary pressure in the liquid rises until it is equal to the solid stress; then menisci recede into the pores and shrinkage stops.

$$C_P = 2\gamma_{LV}\cos(\theta)/r_P \qquad (1)$$

Here θ is the wetting angle, r_P is the pore size, and γ_{LV} is the liquid-vapor surface tension. Because r_p can be of molecular dimensions, the magnitude of the capillary pressure can be enormous (up to 1000 bar), and the capillary pressure represents a strong driving force to densify a drying film. The elastic modulus (E) of the network represents its ability to resist collapse. For both wet and dry silica gels, the following relationship between modulus and porosity (ϕ) is observed [9,11]:

$$E \propto (V_o/V)^{m)} \qquad 2.5 < m < 4 \qquad (2)$$

In Figure (2) the average solid-phase stress and the maximum attainable liquid-phase pressure stress (defined as porosity times the maximum capillary pressure based on the pore radius) are scaled by the initial network modulus (E_0). The maximum liquid-phase pressure rises as the pore radius decreases until the liquid-phase and solid-phase stresses are equal. We call this point the critical point at which shrinkage stops, thereby establishing the final pore size. As shown in Figure (2), for sol-gel systems with low enough modulus or large enough capillary pressure, it is possible to collapse the pores completely because the driving force for network shrinkage (the maximum effective pressure force) can be always greater than the resistance to deformation (solid-phase stress). Of course, as pores collapse to molecular dimensions, the concept of capillary pressure breaks down, requiring additional physics.

5. PREDICTIONS OF STRESS DEVELOPMENT IN DRYING FILMS

After gelation, stresses develop in a shrinking sol-gel coating as it dries. Using the large deformation theory of Segalman et al. (1992)[12] for continuous gels (i.e. not porous), we have predicted the stress evolution for a coating adhering to a deform-

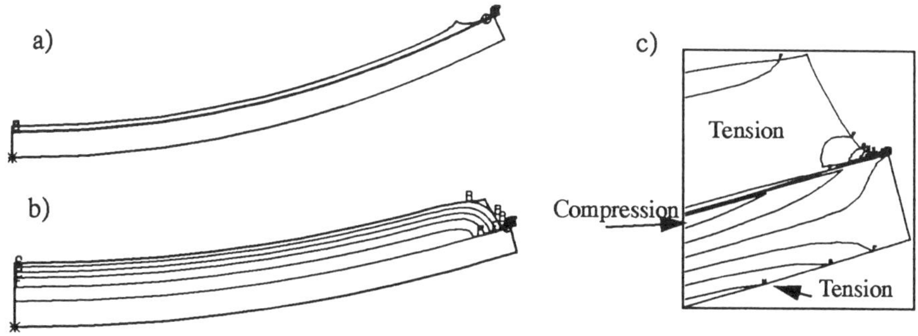

Figure 3) Predictions of shrinkage and deformation in a continuous coating on a stiff, flexible substrate. a) dry curling coating, b) partially dry coating - contours of shrinkage, c) τ_{xx} stress near edge of b) showing coating in tension, and substrate in both compression and tension.

able subtract. In this theory, the stress development and solvent transport are fully coupled, but there is no stress driven diffusion. Typical results are shown in Figure (3); they compare well with predictions in the literature [13,14]. As a coating shrinks the solid matrix is put in tension and the substrate curls towards the coating. At the coating/substrate interface, the substrate is in compression, and the backside of the substrate is in tension.

The largest deformations and stresses in the coating arise near the free edge where the coating is free to deform both in the thickness and substrate directions. In practice coatings produced by the sol-gel technique crack if the film thickness is larger than a certain critical thickness $h_c = 0.5 - 1\ \mu m$. For films that adhere well to the substrate, the critical thickness for crack propagation or the growth of pinholes is given by [15]:

$$h_c = (K_{Ic}/\sigma W)^2 \qquad (3)$$

where K_{Ic} is the critical stress intensity and W is a function that depends on the ratio of the elastic modulus of the film and substrate (for gel films W = 1). In films thinner than h_c, the energy required to extend the crack is greater than the energy relieved by crack growth, so cracking is not observed. The theory shown here for predicting stress development due to coating shrinkage could be used to estimate the stress, σ, in equation (3) and predict when coatings may crack.

6. REFERENCES

1. Brinker, C. J. and Scherer, G. W., *Sol-Gel Science*, Academic Press, Inc., San Diego, 1990.
2. Brinker, C. J., Hurd, A. J. Frye, G. C., Schunk, P. R. and Ashley, C. S. "Review of Sol-gel thin film formation", *Journal of Non-Crystalline Solids*, **147&148**, 424-436 (1992).

3. Walker, J., "The Amateur Scientist: What causes the 'tears' that form on the inside of a glass of wine?", *Scientific American*, 161-169 (1983).

4. Leenaars, A.F.M., Huethorst, J.A.M., and Van Oekel, J.J., "Marangoni Drying: A New Clean Drying Process", *Langmuir*, **6**, 1701, 1703, (1990).

5. Hurd A.J. and Brinker, C.J., "Ellipsometric Imaging of Drying Sol-Gel Films", *Materials Research Society Symposium Proceedings*, 121, 731-743 (1988)

6. Schunk, P. R. and Rao, R.R., "Finite Element Analysis of Multicomponent Two-Phase Flows with Interphase Mass and Momentum Transport", *Internationa. Journal for Numerical Methods in Fluids*, (1994).

7. Cairncross, R. A., Francis, L. F., and Scriven, L. E., "Predicting Drying in Coatings that React and Gel: Drying Regime Maps", *AIChE Journal*, (1995).

8. Kay, B.D. and Assink, R.A., "Sol Gel Kinetics: II. Chemical Cessation Modeling", *Journal of Non-Crystalline Solids*, **104**, 112-122 (1988)

9. Smith, D.M, Scherer, G.W, and Anderson, J.M., "Shrinkage During Drying of Silica Gel", accepted in *Journal of Non-Crystalline Solids* (1995).

10. Chen, K.S. and Schunk, P.R., "Modeling of Drying, Stress Development, and Pore Structure Evolution in Membrane Thin-Film Fabrications", presented at the *6th Annual Joint meeting of the New Mexico Sections of the American Ceramic Society and Materials Research Society*, Albuquerque, NM (1994).

11. Scherer, G.W., *Journal of Non-Crystalline Solids*, 109 (1989) 183.

12. Segalman, D., Witkowski, W., Adolf, D., and Shahinpoor, M., "Theory and Application of Electrically Controlled Polymeric Gels", *Smart Mater. Struct.*, **1**, 95-100, (1992).

13. Tam, S-Y, Stolarski, H.K, and Scriven, L.E., "Stress Effects in Drying of Coatings: An Elasto-ViscoPlastic Model", presentation at *AIChE National Meeting*, Atlanta, GA (1994).

14. Powell, R.W., Kevrekidis, I.G., and Christodoulou, K.N., "Simulation of Simultaneous Transport and Hardening in Polymer Gels", presentation at *AIChE National Meeting*, San Francisco, CA (1994).

15. Evans, A.G., Drory, M.D., and Hu, M.S., "The Cracking and Decohesion of Thin Films", *Journal of Materials Research*, **3**, 1043-1049 (1988).

MELTING AND EVAPORATION OF FLUORIDE NANO-CLUSTERS

A. Dornford-Smith, V. L. Bulatov and Robin W. Grimes
Department of Materials, Imperial College, London SW 7 2BP, U.K.

ABSTRACT

Molecular dynamics calculations have been used to simulate the high temperature behaviour of NaF, CaF_2 and LaF_3. Of particular note are the congruous melting of Na and F in NaF, the enhanced mobility of F over Ca near the melting point of CaF_2 and the high mobility of F within a static lattice of La at least 200 K below the melting point of LaF_3.

INTRODUCTION

It is now possible to routinely predict certain quite complex properties of defective and non-defective ceramic materials using atomistic simulation techniques. Often such methods are based on an ionic description of the lattice; the interactions between atoms are modelled in terms of Coulombic forces with additional effective short-range parameterized potentials. Indeed, a great number of problems have been addressed using this force description, most often within the static lattice quasi-harmonic approximation [1]. This has been facilitated by the fact that such static lattice calculations can easily predict lattice properties such as bond lengths, unit cell volume, elastic constants etc. By comparing these to experimental values it is possible to develop successful procedures by which the short-range inter-atomic potentials can be generated [2]. As a consequence, potential parameter sets are also available for use by methods based on molecular dynamics (MD).

In most cases, MD procedures use periodic boundary conditions in order to simulate enough ions to make the simulation meaningful. Clearly this restricts the application of MD to problems in which a free surface is not important. Fortunately with the great increase in both computational processing power and memory availability, it is now possible to model explicitly sufficient ions that nano particle sized materials are within sight. As a consequence we have embarked on a program of work that aims to model particles of the order of 10^3-10^4 ions, either in vacuo or on substrates [3]. However, here we report results concerning more modest clusters of a few hundred ions.

The materials of concern are NaF, CaF_2 and LaF_3. Throughout this series we assume a full ionic description of forces, i.e. that the ions are F^-, Na^+, Ca^{2+} and La^{3+}. The short-range forces between ions are modelled using two-body inter-ionic potentials of the form;

$$E(r) = Ae^{-r/\rho} - C/r^6 + C/r^{12}$$

where A, ρ and C are adjustable parameters. The parameters were chosen with reference to the corresponding bulk materials and in the case of NaF and CaF_2, where the data was available, the lowest stoichiometric molecules. The potential parameters used in these studies are reported in Table I. The contention is; if the experimental data concerning the molecules and bulk solid are reproduced successfully, the parameters should be suitable to model any sized stoichiometric cluster in-between these two extremes. It might be noted that in choosing these particular fluorides we avoided issues concerning changes in ionicity with cluster size since the smallest molecular species are known to be highly ionic.

METHODOLOGY

In MD we iteratively solve Newton's equations of motion for a set of ions. The MD algorithm predicts the positions of the ions (and often other variables such as velocities and accelerations) at a time $\{t + \delta t\}$ given a knowledge of the positions etc. at time t. Since the aim is to model both melting and evaporation of fluorides, this places a great strain on the computational technique. In particular, as the MD simulation proceeds, we must ensure that energy is conserved effectively. Consequently we have chosen to employ a Gear 5th order predictor corrector algorithm [4]. This is known to be a particularly robust technique. Nevertheless, since we must be prepared to model a range of temperatures from 0K to over 2000K, the problem of choosing a suitable time step, δt, is paramount. The choice of time step should be such that ions do not move too far during δt

Table I. Short-range potential parameters

NaF Potentials	A (eV)	r (Å^{-1})	C (eV Å^6)
F–F	911.69	0.2707	13.8
Na–F	1459.76	0.2605	0.0
Na–Na	1788.18	0.1586	0.0
CaF_2 Potentials			
F–F	1127.7	0.2753	15.83
Ca–F	1273.8	0.2997	0.0
Ca–Ca	44898.1	0.1690	0.0
LaF_3 Potentials			
F–F	1127.7	0.2753	15.83
La–F	1457.8	0.3257	0.0

or the prediction will cease to be sufficiently accurate. On the other hand, if the time step is too small, the time evolution of the simulation will be overly restricted. The problem is more acute since at low temperatures ions move relatively much slower compared to higher temperatures. The solution to this dilemma is to use variable time steps; longer time steps at low temperatures and slower time steps at higher temperatures. The code *Penicillin* [5], is able to choose the time step automatically by testing for energy conservation and adjusting the time step as the simulation proceeds. In all cases, the heating rate was 1 degree every 10^{-14} seconds.

RESULTS

The analysis of results is based on monitoring, as the simulation proceeds, the change in; a). the kinetic temperature and b). the root mean square displacements (RMSD) of ions (defined as $\sqrt{\langle(\Delta r)^2\rangle}$ where Δr is the extent of the movement of an ion from its original position. In what follows we shall compare the RMSD's of the anions and cations in each material. Details concerning the temperature behaviour of NaF and CaF_2 clusters is reported in [3]. Note that although here we discuss results for specific clusters, the behaviours described are very similar to those found for other similar sized clusters.

The 216 ion NaF cluster.

As the cluster temperature is increased, the RMSD shows a slow steady increase indicative of the thermal expansion behaviour expected of a simple ionic solid. Upon melting, there is a sharp increase in slope commensurate with the enhanced mobility of ions in a liquid (see Figure

1). The suddenness of the increase in slope is due to the melting occurring over only a few time steps. Although in Figure 1 the RMSD's for both Na and F are shown, the contours are almost indistinguishable over this temperature range and beyond, suggesting that the general atomic trajectories of the two species are very similar. All these features were confirmed by animating the clusters using molecular graphics techniques [3]. As the temperature of the simulation was increased further, a dynamic loop-like structure was formed (as discussed in detail previously [3]). This allowed the formation of small clusters of 6, 8 and occasionally 10 ions which were sometimes able to escape the main body of the cluster. Material was also lost through the ejection of NaF dimers and $(NaF)_2$ tetramers. In all cases material was lost in stoichiometric units.

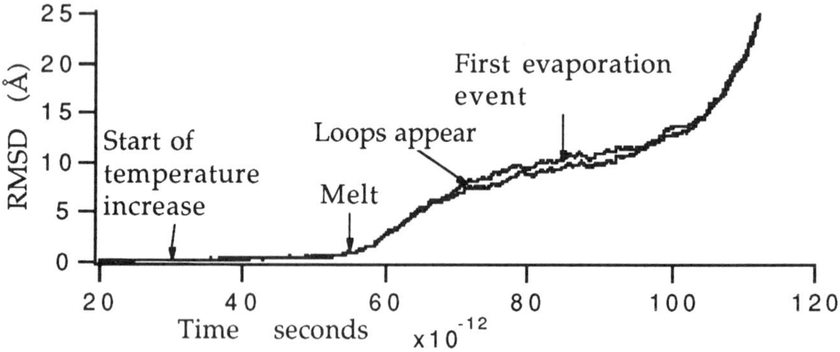

Figure 1. The behaviour of the RMSD's of Na and F ions in 216 ion NaF cluster.

The 312 ion CaF_2 cluster.

The RMSD of the cations shows a slow increase while the cluster is "solid" but increases significantly upon melting, i.e. when ions become free to move throughout the cluster. The change in RMSD slope for the cations is sudden as in NaF. However, the change in slope for the anions is more gradual (see Figure 2). This is due to fluorine ions moving through interstitial sites in the structure at lower temperatures. However, there seems little evidence that the ions occupy such sites for any length of time, i.e. the anion sublattice does not lose its integrity. Such behaviour has also been documented in bulk calculations [3]. At higher temperatures, material evaporates from the main body of the cluster. In contrast to NaF, in almost all cases triatomic CaF_2 molecules are produced. Certainly no loop mechanism is available to aid in cluster formation as in NaF [3].

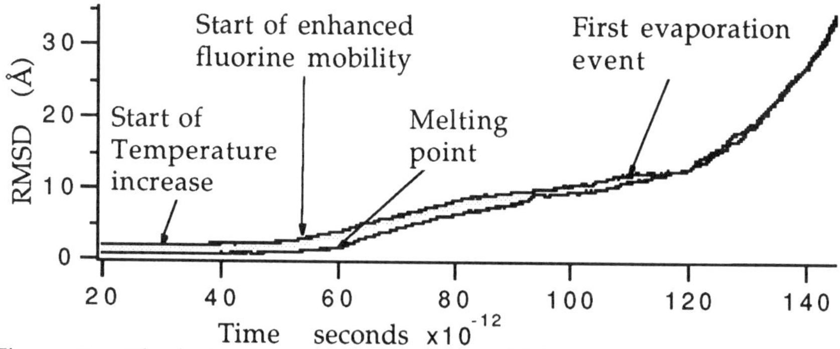

Figure 2. The behaviour of the RMSD's of F (upper curve) and Ca (lower curve) in 312 ion CaF_2 cluster.

The 160 ion LaF_3 cluster.

This cluster behaves similarly to CaF_2. However, rather than enhanced fluorine mobility, the anion sublattice seems to be much more disorder, although the average five-fold coordination around the La sublattice is certainly maintained. Note the much larger time between the onset of anion disorder and melting compared to the equivalent points in CaF_2 (compare Figures 2 and 3). In fact, LaF_3 is known to become a super-ionic conductor above ~1100K [6], essentally at the same temperature at which the fluorine ions show greater mobility. At higher temperatures, material evaporates, in this case mainly in the form of LaF_3 tetramer molecules although larger $(LaF_3)_2$ units are not uncommon. This is due to there being a particularly stable $(LaF_3)_2$ molecular geometry (where the La ions are tetrahedrally coordinated by F, the two tetrahedral sharing one edge i.e. a pair of F ions).

DISCUSSION

These simulations are based on inter-ionic forces which reproduce the bulk and molecular properties of their respective materials. Despite the relatively small size of the clusters, certain features associated with melting and anion disorder reflect well known bulk properties. For example, the melting point of the NaF cluster was 1300K close to the experimental value of 1266K. However, whether this was due to the

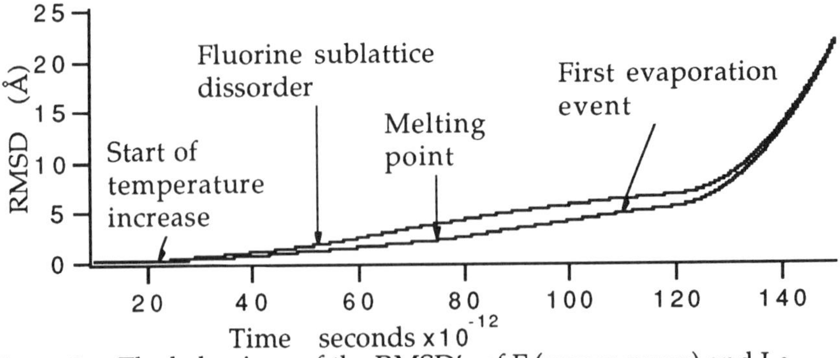

Figure 3. The behaviour of the RMSD's of F (upper curve) and La (lower curve) in 160 ion LaF_3 cluster.

geometric regularity of the cluster (which was a 6x6x6 ion cube) or to how we define temperature must remain in contention until we have carried out more expensive simulation studies. Similarly, note the enhanced fluorine ion mobility up to 150K below the melting point in CaF_2 and the disordering of the fluorine sublattice in LaF_3 close to the experimental super-ionic transition temperature. Nevertheless, it would seem that MD calculations based on nano-sized clusters will be able to provide useful insight into some of the dynamical properties of ceramic materials.

Acknowledgements

We are grateful to the EPSRC for providing funding for V. L. Bulatov under the ROPA scheme, GR/K38014.

REFERENCES

1. Computer simulation of defects in polar solids; special issue of J. Chem. Soc. Faraday Trans., **85**[5] (1989). Guest editors C. R. A. Catlow and A. M. Stoneham.
2. The practical calculation of interionic potentials in solids; special issues of Mol. Simul., **4**[5] and **5**[2] (1990). Guest editors A. H. Harker and R. W. Grimes.
3. A. Dornford-Smith and R. W. Grimes, "Novel vapourisation mechanisms for NaF and CaF_2 nano-crystallites in vacuo," *Phil. Mag.*, in press.
4. M. P. Allen and D. J. Tildesley, "Computer Simulation of Liquids," Clarendon Press, Oxford, 1987.
5. Penicillin if freely distributed by the authors from Imperial College where it is presently being developed.
6. G. A. Jaroszkiewicz and J. H. Strange, J. Phys. C: Solid State Phys., **18**, 2331-49 (1980).

Ceramic Processing

SMOOTHED PARTICLE MODELING FOR CERAMIC INJECTION MOLDING

Takashi Iwai, Tatsuhiko Aizawa and Junji Kihara
Department of Metallurgy, University of Tokyo
7-3-1 Hongo, Bunkyo-ku, Tokyo, Japan

ABSTRACT

To investigate the complex physical flow behavior of the powder-binder compound, and to optimize the process condition in the ceramic injection molding process, a new numerical model has been developed. The smoothed particle modeling is applied in the mesoscopic level, to link both the micro- and the macroscopic physics of mixed compound, as an extension of the granular model which represents the microscopic rheology of mixed compound. This numerical model especially focuses on the prediction of the inner state of mixed compound and the effects of powder characteristics to flowability and formability. Several numerical examples are demonstrated and reveal the relation between the flowability and the microscopic structure in this kind of material.

INTRODUCTION

The powder injection molding process has been noted to be a leading method of the precise and net shape forming of ceramic products. Although a larger quantity of binder is necessary in this process than in other processes to improve the molding performance, the powder properties as particle size, particle size distribution and particle shape are controlled to achieve as high solid loading as possible. Since this kind of mixed compound indicates the strong nonlinear behavior accompanied with agglomeration and segregation of powders and binders, the mixing and molding stages are considered to account for the major part of this process for accuracy and productivity. Nevertheless, relatively few studies have been published on this field [1,2], especially in the numerical simulation, little theoretical consideration has been made for the relationship between powder characteristics and flowability of mixed compounds.

For microscopic evaluation of flowability of the powder-binder mixed compound in the ceramic injection molding process, the metal modeling with composite elements has been developed by authors [3–5]. The granular model traces all the motions of each granule, and determines the deformation of the system by solving all these equations. The interaction between particles and binders are considered as the inter-element forces among composite elements which consists of both a metallic powder core and a binder shell. The model is extended to three dimensional code, and linked with the thermo-mechanical coupling method to represent the irregular rearrangement of particles or the micro-segregation accompanied with consolidation of binder material.

Although the granular model is suitable to show the explicit relation between powder structures and the flowability of the mixed compound, some problems still remain for the practical systems. The microscopic analysis like the granular model needs a huge number of elements, and occasionally causes divergence of computation. The new time integration scheme and algorithm by sub-cycling mechanism shows considerable improvement against the divergence [4], but it is difficult to apply the method directly for the mixed material of minute powders like ceramics due to the sharp and sensitive response.

In this paper, the smoothed particle model is described as a new numerical modeling. This mesoscopic model is developed to resolve their difficulties above in larger systems, and at the same time to consider the relation between macroscopic flow behavior and microscopic physical properties like agglomeration and segregation for the dense ceramic powder-binder mixed compound. Some viscosity tests are employed with use of this method to evaluate and validate the present formulation.

SMOOTHED PARTICLE MODELING

For the powder-binder mixed compound, it is necessary to consider the interaction between particles and also binders. Even by use of the granular model, relatively practical problems have been simulated with more than 10^4 particles. However in the case of ceramic powders, as easily imagined, it is difficult to handle all the powders in the material; in such a material, over-million particles are to be free. Hence, we introduce the super particle element instead of one particle element in the conventional granular model. Fig. 1 depicts the schematic view of the present smoothing method in the powder-binder mixed compound. Powder particles in the mixed compound are assumed to form super particle elements, i.e. powder clusters. In each element, powders are distributed in the given alignment.

The present smoothing method is based on the an wellknown particle method, the smoothed particle hydrodynamics (SPH), which was developed by Gingold and Monaghan [6] originally for astronomic interests. SPH application now spreads over many research fields from plasma to molecular dynamics [7].

The basic concept of SPH is the interpolation by kernel bases. In SPH, a physical

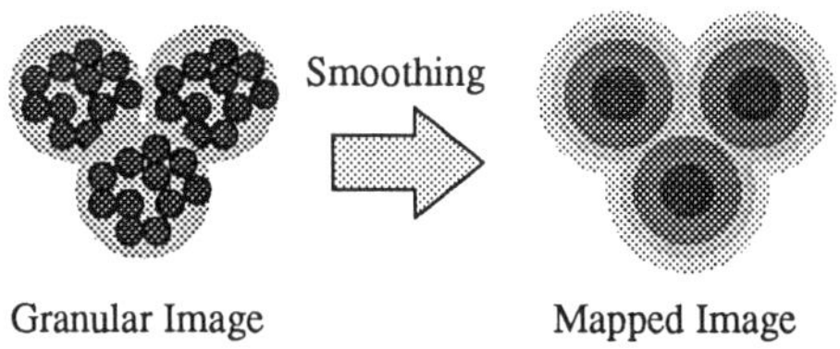

Figure 1: Schematic view of smoothing method in the powder-binder mixed compound

value $f(\boldsymbol{x})$ at point $\boldsymbol{x}$ in material is written by summation of kernel basis of particle i as

$$f(\boldsymbol{x}_i) = \sum_{j \in h_i} f(\boldsymbol{x}_j)\, W(\boldsymbol{x}_i - \boldsymbol{x}_j, h_i)\, \Delta V_j, \tag{1}$$

where W is the kernel function, and h_i is the smoothing length. For simplicity, the kernel $W(\boldsymbol{x}_i - \boldsymbol{x}_j)$ is abbreviated as W_{ij} in the following. V_j represents the specific volume at point j, which is defined by the mass m_i and the density ρ_i as

$$\Delta V_i = m_i/\rho_i, \quad \rho_i = \sum_{j \in h_i} m_j\, W_{ij}. \tag{2}$$

The spatial differential ∇f is given from Green's theorem by

$$\nabla f_i = \sum_{j \in h_i} f_j\, \nabla W_{ij} \Delta V_j. \tag{3}$$

In the smoothed particle modeling, to consider the internal change in the mixed compound in local, the microscopic interaction between powders and binders is decomposed to two types: the contact of powders and viscous friction of powders and binders. The former is approximated by the linear elastic response. The elastic force $\boldsymbol{F}_{ij}$ which takes place by the overlap $\Delta \boldsymbol{u}_{ij}$ of two elements i and j is given by

$$\boldsymbol{F}_{ij} = \boldsymbol{K}_{ij} \Delta \boldsymbol{u}_{ij}. \tag{4}$$

The elastic stress is expressed by the volume average

$$\boldsymbol{s}_i^* = \frac{1}{\Delta V_i} \sum_j \boldsymbol{L}_{ij} \boldsymbol{F}_{ij}, \tag{5}$$

where $\boldsymbol{L}_{ij}$ is the vector to the contact point. To represent the response of the super particle element, we introduce a strain distribution function $\boldsymbol{G}_j$,

$$\boldsymbol{S}_i^* = \frac{1}{\Delta V_i} \sum_j \boldsymbol{G}_j \boldsymbol{L}_{ij} \boldsymbol{F}_{ij}. \tag{6}$$

In the normal situation, a simple Maxwellian type formulation is suitable for this function, given by

$$\boldsymbol{G}_i = \boldsymbol{A}_i \sum_j exp(-0.5\boldsymbol{v}_{ij}) \tag{7}$$

where $\boldsymbol{v}_{ij}$ is a ratio of the variance of the particle stiffness over the averaged stiffness, and $\boldsymbol{A}_i$ is a constant to normalize the function $\boldsymbol{G}$. In the simulation, the function $\boldsymbol{G}$ is given before applying the smoothing method as a prescribed form computed from the randomly sampled particle cluster.

The latter, the viscous term between powders and binders, is approximated from the integration of the interaction around the powders. The stress on a powder particle acts around its surface, then it is necessary to integrate over the particle surface area. When the viscous interaction f is defined on the surface of a powder S, the nodal value is described as

$$f = \sum_j \left[\int_S f(r_j) W_j(r_j) dr \right] = \sum_j \bar{W}_j f_j \Delta V_j, \tag{8}$$

where $\bar{W}$ represents the averaged kernel on the surface S. The averaged kernel can be calculated from the expansion of W as

$$W(r) = W_o + r_k \frac{\partial W_o}{\partial r_k} + \frac{1}{2} r_k r_l \frac{\partial^2 W_o}{\partial r_k \partial r_l} + \cdots, \tag{9}$$

then becomes a simple form

$$\bar{W} = W_o + \frac{r^2}{6} \nabla^2 W_o, \tag{10}$$

where W_o means the kernel basis on the nodal point. Again in the super particle element, the averaged kernel becomes a overall kernel by using the distribution function $\boldsymbol{H}$, which is prescribed as well as in the strain distribution:

$$f = \sum_j \boldsymbol{H}_{ij} \bar{W}_j f_j \Delta V_j. \tag{11}$$

NUMERICAL EXAMPLES

To validate the present modeling, a simple shear deformation is tested. Two parallel plates move for each opposite horizontal direction on the computational condition listed in Table 1. Here, three types of powders are employed to investigate the effect of the microscopic parameter to the macroscopic viscous flow. The powders are placed randomly in the flow. Also the uniform placement system is calculated to compare the effect of agglomeration to the flow viscosity. From the given boundary velocity and the calculated stress on boundary walls, the apparent viscosity is estimated as well as in the parallel viscometer used in many experiments.

Table 1: Computational condition of parallel shear flow test.

Powder Material	Alumina
Binder Material	Polyethylene
Powder diameter	1, 5, 20μm
Smoothing length	5 times of diameter
Powder volume	$\sim$40%
Shear strain rate	10$\sim$200 s^{-1}

Fig. 2 shows the change of calculated relative viscosity over the shear strain rate for the various powder diameters. Almost linear relation can be seen in this logarithmic plot as measured in the experimental researches [2]. The similar result was predicted by the microscopic modeling in the case of the metallic powder and the binder mixed system employed in the metal injection molding [3]. In fig. 2, the gradient increases when the powder size decreases. Not only the increase of inter-particle forces with rise of particle surface area but also the agglomeration of the particles result in this drastic increase of the nominal viscosity in the flow.

Fig. 3 shows the effect of agglomeration to the nominal viscosity. The y-axis shows the normalized viscosity by dividing the viscosity η by the uniform viscosity η_u corresponding to the non-agglomerated system. Again in this figure, the compound with smaller powders shows the higher viscosity, while its gradient is more gentle than in fig. 2. The result implies the agglomeration triggers the major increase of viscosity of the mixed compound.

CONCLUSION

To describe the mechanical behavior of dense and viscous powder-binder mixture materials in the injection molding process, a novel numerical modeling was proposed. This smoothed particle model was developed to resolve the problem occuring in the our previous microscopic granular model.

The parallel shear flow was tested to evaluate the flowability of such compound material. The good logarithmic relation was calculated over various powder diameters as known in many measurements for this kind of material. Also nonlinear curves were found in the normalized viscosity against the the non-agglomerated system. These results show the validity of this model as a prediction method for the microscopic state of the mixed compound. Since the proper constitutive relationship for such a material is still unknown, these direct descriptions of powders and binders must be inevitably necessary to recognize the physical experimental phenomena quantitatively.

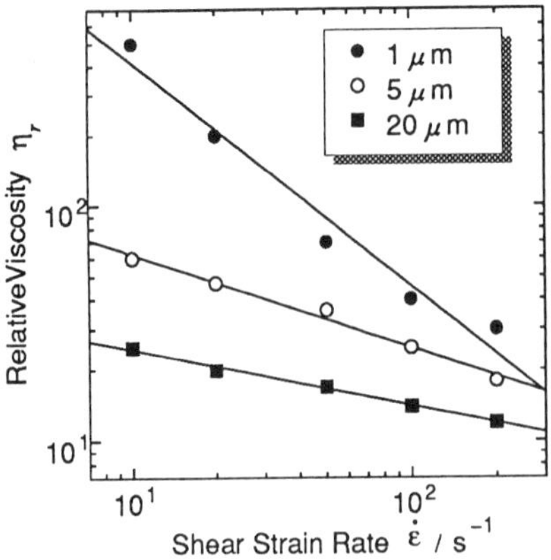

Figure 2: Calculated viscosity over the shear strain rate in the various powder diameters.

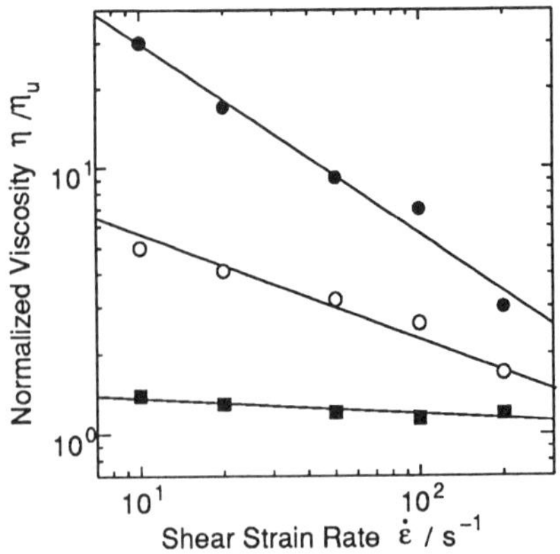

Figure 3: Calculated normalized viscosity in the various powder diameters.

REFERENCES

[1] J.H. Dow, M.D. Sacks and A.V. Shenoy, "Dispersion of Ceramic Particle in Polymer Melts", Ceramic Transactions, Vol. 1, Part A, Ceramic powder Science, ACerS, 1988, pp. 380-385.

[2] K. Okada, H. Kurisaka and Y. Akagi, "Dispersion of Powders in Ceramic Injection Molding Mixtures", Proc. Sixth Int. Symp. on Agglomeration, Nagoya, Japan, 1993, pp. 687-691.

[3] T. Aizawa, T. Iwai and J. Kihara, "Granular Modeling of Steel Powder Flow and Compaction in Injection Molding", Proc. Powder Injection Molding Symp., MPIF, 1992, pp. 419-433.

[4] T. Iwai, T. Aizawa and J. Kihara, "Granular Modeling for Dense Powder-Binder Compound Flow in Metal Injection Molding", J. Fac. Eng., Univ. Tokyo B, Vol. XLII, No. 4, 1994, pp. 317-347.

[5] T. Iwai, T. Aizawa and J. Kihara, "Three Dimensional Granular Modeling for Metal Injection Molding", Proc. 1993 Powder Metallurgy World Congress, Part 1, JSPPM, 1993, pp. 222-225.

[6] R. Gingold and J.J. Monaghan, "Smoothed Particle Hydrodynamics: Theory and Application to Non-Spherical Stars", Mon. Not. Roy. Astron. Soc., Vol. 181, 1977, pp. 375-389.

[7] W.G. Hoover and C.G. Hoover, "Smoothed-Particle Hydrodynamics and Nonequilibrium Molecular Dynamics", AMD-Vol. 180 / PDP-Vol. 268, ASME, 1993, pp. 231-242.

THREE DIMENSIONAL FINITE ELEMENT PROCESS SIMULATION FOR GREEN MATERIALS WITH ELEMENT CONTROL

Tatsuhiko Aizawa, Jan Terziyski and Junji Kihara
Department of Metallurgy, University of Tokyo
7-3-1 Hongo, Bunkyo-ku, Tokyo 113, Japan

ABSTRACT

Three dimensional finite element process simulator was developed to describe the rigid-plastic and visco-plastic behaviors of green materials. Different from the conventional simulation codes, our developed system has a function of element control not only to make automatic remeshing and remapping for super-large deformation and flow but also to perform arbitrary lagrangian formulation for direct consideration of tool-work interactions. Forward extrusion process is taken for a numerical example to demonstrate the validity and effectiveness of our developed approach.

INTRODUCTION

Three dimensional process simulators can be available to actual process design analysis owing to rapid advancement of computer technology; ABAQUS [1] and NIKE [2] are typical commercial programs with a function of nonlinear analysis. Since super-large deformation and flow of materials must be dealt with in actual process design, reliable and accurate formulation with computational modeling should be equipped with a promising simulator: Robust element design against local geometric distortion, Accurate treatment for tool-work interaction, and Automatic control of computational model. Since the incompressibility condition is imposed on the velocity field for rigid-plastically or visco-plastically deforming green materials, adaptive element design must be performed to trace large flow and circulation of green body even in local. As discussed in Refs. [3,4,5], a composite element was newly developed as a robust finite element scheme adaptive to both two and three dimensional hyperelastic, rigid-plastic and elasto-plastic analyses. This composite element with 10 nodes is composed of eight sub-elements, and both velocity and pressure are independently approximated by piecewise linear function in each sub-element and constant in a parent element, respectively. Refs. [5], [6] and [7] demonstrated the accuracy and reliability of this element.

Remeshing and remapping procedure is an essential element control to continue computations up to the prescribed configuration. Computational model is successively updated to keep the accuracy of numerical solutions within a tolerance; models must be automatically controlled in the adaptive manner to reduction of various errors and geometric distortions. As have been studied in Refs. [8], [9], [10], [11] and [12], computational geometry provides rational perspective to make adaptive element control. However, tool-work interaction or Euler and Lagrangian coupling can never be dealt with satisfactorily only by mesh refinement nor adaptive control. As partially discussed in

Refs. [13,14], Arbitrary Lagrangian Eulerian (ALE) method must be one of the most promising methods to evaluate the three dimensional contact behaviors; since every physical quantities are measured in the reference frame which is different from the spatially fixed coordinates and the material ones, actual material flow at the vicinity of die corner can be accurately described together with accurate prediction of contact surface and contact tractions.

In the present study, authors are only concerned with the ALE formulation and related element control, which is indispensable to simulate the deformation and flow of green materials in forward extrusion. Essential fundamentals for ALE formulation are briefly introduced with some comments on the element control necessary for direct consideration of tool-work interactions. Forward extrusion process is taken for an example to make comparison between Lagrangian model and ALE model. Through discussions, the validity and effectiveness of our approach is demonstrated for reliable process simulation of green compound.

ALE-FORMULATION FOR ELEMENT CONTROL

The updated Lagrangian or total Lagrangian approaches have been and still are employed in the conventional programs; hence, they have to suffer from two major deficiencies. One is the inevitable remeshing and remapping to diminish geometric deterioration of models and to trace time history of physical quantities. The other is inaccurate consideration of tool-work interactions without increase of number of elements; e.g. a die corner should be a singular point where material velocity must be different from mesh velocity, and very fine meshing is necessary to reduce errors when this singular point is dealt with by a regular node.

In order to overcome these two demerits, element control must be taken into account of formulation to deduce reliable and accurate computational modeling. Two approaches for this element control are considered in the present study; adaptive and automatic remeshing/remapping procedure, and Arbitrary Lagrangian and Eulerian (ALE) method. In the former, three dimensional computational geometry was used to construct automatic element controller with both geometric distortion of composite sub-elements and deviation of equivalent strain-rate employed as an error indicator for each element [6,11]. This procedure works well to cope with very large deformation and flow of green materials in the absence of singular die corners [12].

In the ALE description, any physical quantity f is measured in the specified reference frame, which is not necessary coincident either with the spatial coordinates or the material coordinates. As shown in Fig. 1, actual material velocity u is represented as a sum of the velocity w of moving reference frame χ and relative material velocity v measured from the reference frame. Since these coordinate systems change with time, the Jacobian J, or mapping function from reference frame to the laboratory coordinates, must also vary with time; evolution of J should be described by

$$\partial J / \partial t = \det(J) \nabla w , \tag{1}$$

where det(J) denotes the determinant of Jacobian J. Remember that any physical quantity must be indifferent to selection of coordinate systems to be used: i.e. physical quantity $F(\chi,t)$ defined in the reference frame should be identical to $f(x,t)$. Hence, time evolution of physical quantity can be represented even in the reference frame:

$$\partial[Jf(x,t)]/\partial t = J \{\partial F(\chi,t)/\partial t + \nabla[F(\chi,t)\cdot w]\} . \tag{2}$$

Equation (2) becomes a basic relation between the time derivatives of a function both in the reference and laboratory coordinate systems, which should transform any fundamental equations from one analytical frame to the other.

Both inner and external work powers are chosen as f in the above. Equations (1) and (2) results in the variational equations in the reference frame. By using the inverse Jacobian mapping, this variational equations are further rewritten into a ALE-variational equation. Different from the conventional equations in the Lagrangian frame, advective term of a power must be included:

$$\int_V W_{int}(V_\chi, u, t)\, dV + \int_{Sn} W_{int}(V_\chi, u, t)\, w_n\, dS = \int_{S\sigma} W_{ext}(V_\chi, u, t)\, dS. \qquad (3)$$

Here W_{int} and W_{ext} are internal and external work power, respectively, and w_n the normal component of reference frame velocity w across the interface of ALE-region.

Let us consider an adaptive computational model to this ALE formulation in the forward extrusion process. As depicted in Fig. 2, the original material is represented by three regions in general: Lagrangian region for materials with less coupling with tools, standard ALE region with contact state with tools and ALE-Eulerian region with singular nodes and edges. Owing to the above formulation, advective terms might be considered on the interface S_n among these regions and boundaries of materials.

Direct discretization and linearization during the prescribed time increment enables us to deduce the incremental finite element equations corresponding to equation (3). Nonlinear constitutive equations are employed in computation: e.g. equivalent stress is given by $\sigma_e = \kappa \cdot \varepsilon_e^n \cdot \varepsilon_e^m$ for the effective strain ε_e and strain rate ε_e. In actual computation, iteration scheme with use of linearized scheme is utilized to obtain the convergent solution, and time integration advances by variable increments.

NUMERICAL EXAMPLES

Forward extrusion process is typical but fundamental system widely used to yield thin ceramic plates and films. In this process design, two important issues need to be solved by three dimensional numerical simulator: 1) Optimum or rational determination of die cavity geometry from inlet to outlet, and 2) Accurate prediction of varying work material geometry and contact force distribution during extrusion. On the basis of these informations, a database can be constructed to estimate die shape and allowable accuracy of worked green plates and sheets for each event of processing.

Die configuration is depicted in Fig. 3 together with the initial finite element model. Even at the absence of friction on the tool surface, since tool surfaces are bent at the inlet, specification of boundary conditions becomes difficult in the conventional or the Lagrangian approach. Here, nodes on the edge like a point B are assumed to move along the edge in the Lagrangian analysis. To be noted, no assumptions are necessary for ALE-formulation and related computational modeling; tool-work interaction can be directly dealt with as the coupling between tool model (Eulerian) and ALE-based model. Let us compare the geometric configuration change between the Lagrangian model and the ALE-model. As shown in Fig. 4, artificial diffusion takes place not only along edges but also in the middle of tool surface in the Lagrangian model. Furthermore, apparent high pressure distribution can be seen at the inlet, which is not true to actual processing situation. Due to large geometric distortion, Lagrangian model failed to continue computation at the reduction of r = 20 %. In the ALE-model, no artificial diffusivity nor false high pressure distribution can be seen, and simulation can be continued to the reduction of 40 % or more. Then, ALE-formulation and ALE-modeling must be indispensable to make reliable forward extrusion process of greens and to provide required informations for process design analysis.

CONCLUSION

Element control with the Arbitrary Lagrangian Eulerian method should be effective to consider the interaction between tools/dies and work materials. In the present study, the reference coordinate frame was controlled by the prescribed procedure as the initial condition. In order to make thorough simulations in actual process, however, automatic control with remeshing is indispensable to diminish deterioration of mapping function from the reference coordinates to the materials coordinates [14]. In order to produce annular-shaped and multi-perforated green bodies, a process simulator must consider separation of one green billet into two and more channels and recombination of these fragmented billets into one green body. As discussed in Refs. [15,16,17], advanced Eulerian approach with controlled ALE stepping must be adaptive to describe discontinuity and re-continuity of green material flows in simulation. The interesting point to be noted is that mixing and homogenization processes should be also described by using the multi-material modeling.

Only rigid-plastic and visco-plastic models were employed here as the constitutive equations. In actual process design, various rheological models must be taken into account to deduce the most suitable constitutive relation among them. As developed and discussed in Refs. [18,19], the granular modeling, where a powder particle or a powder compound is represented by one distinct element, is found to be effective to precise discussion of rheological behaviors even in local and to construct an averaged constitutive relation to be used in the present approach.

REFERENCES

[1] ABAQUS theoretical manual, HKS Inc. (1987).

[2] J. Hallquist, NIKE3D: An Explicit Finite-Deformation, Finite-Element Code for Analysis of Static and Dynamic Response of Three Dimensional Solids, LLNL, UCID-18822 (1991).

[3] T. Aizawa and J. Kihara, Rigid-Plastic Finite Element Simulation with Adaptive Element Control for Practical Forging Process, Proc. IUTAM Symp. (1991) 449-460.

[4] T. Aizawa and J. Kihara, Composite Finite Element Elements for Rigid-Plastic Analysis, J. Mat. Proc. Tech. 27 (1991) 55-71.

[5] T.J. Hughes, The Finite Element Method (1989) McGraw-Hill.

[6] J. Terziyski, PhD Thesis of University of Tokyo (1995,March).

[7] T. Yamada and F. Kikuchi, A Mixed Finite Element Method for Large Deformation Analysis of Incompressible Hyperelastic Materials, Theor. Appl. Mech. 39 (1990) 61-73.

[8] T. Aizawa and Y. Yoshino, Three Dimensional Mesh Generation for FEM-Automation, Proc. Post-Conference Seminar of ICES-88 (1988) 29-58.

[9] T. Aizawa, Automatic Generation of Finite Element Modeling for Integrated CAD and CAE, CAD/CAM robotics and Factories of the Future '90 (1990) 273-278.

[10] T. Aizawa, Three Dimensional Triangulation for Forging and Solidification Designs, ASME, PVP-22 (1992) 1-25.

[11] T. Aizawa, Finite Element Control on the Basis of Computational Geometry, Proc. Symp. Computer Aided Innovation of New Materials (1991, Tokyo) 603-608.

[12] T. Aizawa, Finite Element Control and Management for Metal Forming Process Simulation, Proc. Comp. Mech. Symp. (1991, Merborn) 13-16.

[13] T. Yamada and F. Kikuchi, An Arbitrary Lagrangian-Eulerian Finite Element Method for Incompressible Hyperelasticity, Comp. Meth. Appl. Mech. Engg. 102 (1993) 149-177.

[14] J. Terziyski, T. Aizawa and J. Kihara, Three Dimensional Forging Simulation by Rigid-Plastic Analysis with Element Control, Third US National Congress on Computational Mechanics (1995, Dallas).

[15] T. Aizawa, D. Benson and J. Kihara, Powder Deformation, Flow and Fracture Processes in Dynamic Plasticity, Proc. Plasticity-95 (1995, July, Sakai) (to be published).

[16] T. Aizawa, D. Benson and J. Kihara, Direct Simulation of Powder Mixing and Refinement Process in the Mechanical Alloying, Proc. ICES-95 (1995, July, Hawaii) (to be published).

[17] T. Aizawa, J. Terziyski and J. Kihara, Adaptive Physical and Computational Modeling Design to Working and Forming Analyses of Materials, J. Faculty of Engineering, University of Tokyo (B), XLIV (1995) (to be published).

[18] T. Iwai, T. Aizawa and J. Kihara, Granular Modeling for Dense Powder-Binder Compound Flow in Metal Injection Molding, J. Faculty of Engineering, University of Tokyo (B) XLII (1994) 317-347.

[19] T. Aizawa, Granular Modeling for Computer Aided Design of Colloidal Processing, American Ceramic Society (1995) (in Press).

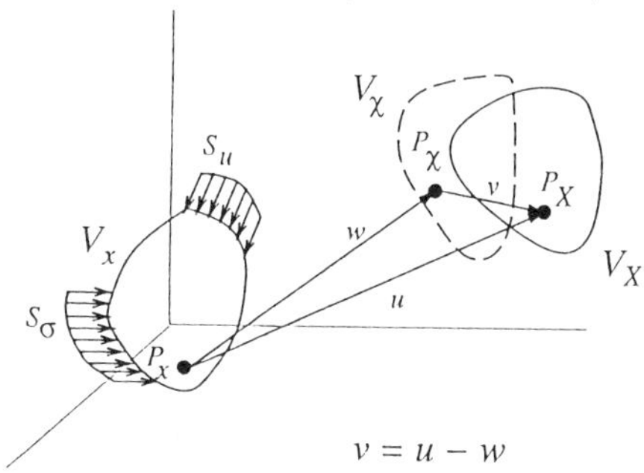

Fig. 1 Kinematic relations in the ALE-formulation.

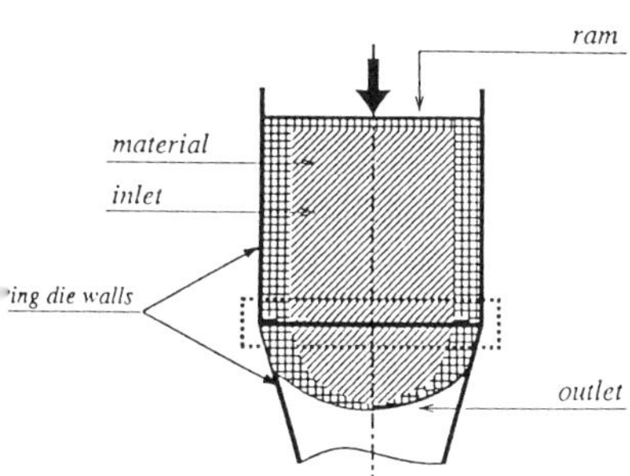

Region	Description	Numerical Treatment	Zone
(hatched)	Lagrangian	The mesh velocity is equal to material velocity. No convective terms are calculated	Internal domains far enough from the die
(grid)	ALE	Each node has two DOFs: mesh and material velocity vectors. The mesh velocity is prescribed by using boundary surfaces normal vectors and volumetric constraints. The convective integrlas are calculated only for the non-constrained corresponding materal DOFs.	All boundary regions and model front tip
(dotted)	ALE-Eulerian	In ALE description, the mesh nodes are fixed on die corners and only the convective terms for the corresponding material DOFs must be calculated. In simulation, the referential nodes can be replaced by other ones stepwise.	Boundary nodes mapped on more than one tool surface

Fig. 2 ALE- models for practical application to forward extrusion process simulation.

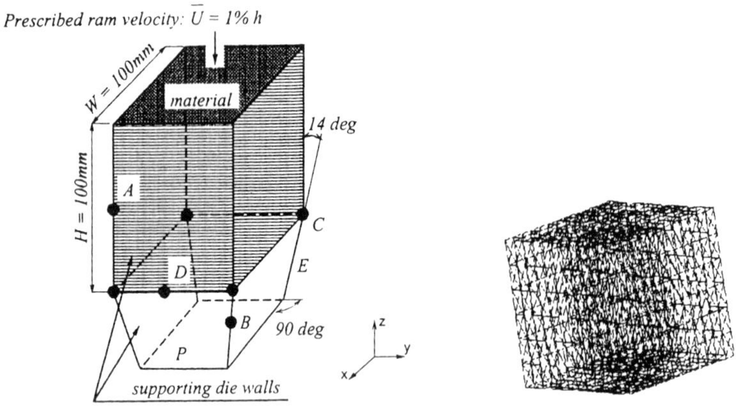

Outline of the process setup

Initial finite element model: 1372 nodes, 648 elements

Fig. 3 Tool configuration in the forward extrusion process with a finite element model.

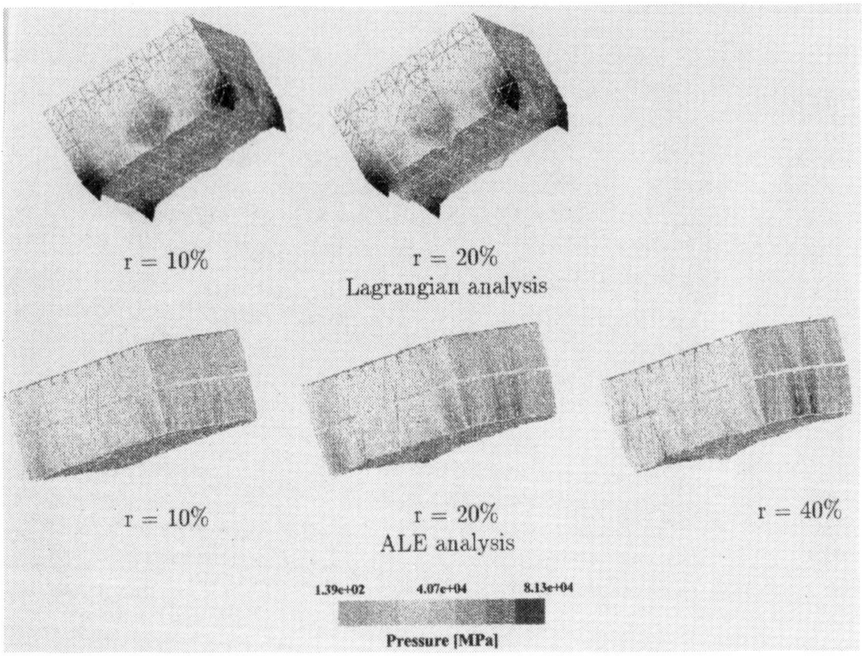

Fig. 4 Comparison of geometric configuration change of work green materials between the Lagrangian model and ALE-model.

Microstructural Evolution and Grain Growth

A KINETIC DIFFUSE-INTERFACE MODEL FOR MICROSTRUCTURAL EVOLUTION IN MULTIPHASE POLYCRYSTALLINE MATERIALS

Danan Fan and Long-Qing Chen
Department of Materials Science and Engineering,
The Pennsylvania State University,
University Park, PA 16802

ABSTRACT

A diffuse-interface model is proposed for studying the kinetics of microstructural evolution in multiphase systems. In this model, an arbitrary multiphase microstructure is described by field variables which take different values in different phases and within different grains, and are continuous across the interphase boundaries and grain boundaries. The kinetics of microstructural evolution is described by the temporal evolution of these field variables obtained through a numerical simulation of the time-dependent Ginzburg-Landau equations. Preliminary results on the microstructures and their evolution in a two-phase solid and solid-liquid two-phase mixtures will be presented.

INTRODUCTION

A general multiphase polycrystalline microstructure contains both interphase boundaries between different phases and grain boundaries between grains of the same phase. Therefore, at a given temperature, it is expected both kinds of interfaces will move to reduce the total grain boundary and interfacial energy. The driving force for a given grain boundary to move is the difference in the chemical potentials of an atom on the opposite sides of the grain boundary. The velocity of boundary migration is determined by the rate at which atoms jump across the boundary. The typical diffusion distance for atoms involved in the grain boundary migration is, therefore, on the order of the grain boundary width which, for a pure material, is about two or three lattice parameters. On the other hand, The driving force for an interphase boundary to move is the difference in the chemical potentials of atoms in large and small second phase particles, which results in a difference in the chemical potentials, and thus the compositions, of solute atoms in the matrix immediately outside the second phase particles. In order for a large particle to grow, however, solute atoms outside a small particle have to diffuse through the matrix grains to regions near the large particle. Therefore, the typical diffusion distance for atoms involved in the interphase boundary migration is in the order of the separation distance between second-phase particles. Due to the different characteristics between grain growth and Ostwald ripening discussed above, essentially all previous theoretical studies treated them as two separate processes. Developing a unified analytical model using the conventional sharp-interface model for simultaneous migration of interphase and grain boundaries faces almost insurmountable difficulties since their migration kinetics depend on many materials parameters such as grain boundary and interphase boundary energies, lattice, grain boundary and interphase boundary diffusivities, and more importantly, the topology of the a multi-phase microstructure.

The main purpose of this paper is to report a new model for realistically simulating the kinetic processes which involve simultaneous migration of interphase and grain boundaries. It is based on our recent computer simulation model for grain growth in single-phase systems [1-2] and coupled grain growth and Ostwald ripening in two-phase systems [3]. In particular, we will present our preliminary results on the application of this model to the kinetics of Ostwald ripening in solid-liquid two-phase mixtures.

KINETIC MODEL

(1) Kinetic model for a Two-Phase Mixture

To describe an arbitrary two-phase (α and β) polycrystalline microstructure using the diffuse-interface theory, a set of continuous field variables were defined,

$$\eta_1^\alpha(r), \eta_2^\alpha(r), \ldots, \eta_p^\alpha(r), \eta_1^\beta(r), \eta_2^\beta(r), \ldots, \eta_q^\beta(r), C(r), \tag{1}$$

where η_i^α (i = 1, ..., p) and η_j^β (j = 1,..., q) are called orientation field variables and p and q are the number of orientations. $C(r)$ is the composition field which takes the value of C_α within an α grain and C_β within a β grain.

Similar to the free energy description by Cahn and Hilliard for interphase interfaces [4], the total free energy of a two-phase system, F, can be written as

$$F = \int \Big[f_o\Big(C(r); \eta_1^\alpha(r), \eta_2^\alpha(r), \ldots, \eta_p^\alpha(r); \eta_1^\beta(r), \eta_2^\beta(r), \ldots, \eta_q^\beta(r)\Big) + \frac{\kappa_C}{2}(\nabla C(r))^2 + \sum_{i=1}^{p} \frac{\kappa_i^\alpha}{2}\Big(\nabla \eta_i^\alpha(r)\Big)^2 + \sum_{i=1}^{q} \frac{\kappa_i^\beta}{2}\Big(\nabla \eta_i^\beta(r)\Big)^2 \Big] d^3r \tag{2}$$

where κ_C, κ_i^α and κ_i^β are the gradient energy coefficients for the composition field and orientation fields, and p and q represent the number of orientation field variables for the α and β phases.

Following the diffuse-interface theory, the energy of a planar grain boundary, σ_{gb}^α, between an α-grain of orientation 1 and another α-grain of orientation 2 may be calculated as follows,

$$\sigma_{gb}^{\alpha} = \int_{-\infty}^{+\infty} \left[\Delta f\left(\eta_1^{\alpha}, \eta_2^{\alpha}, C\right) + \frac{\kappa_C}{2}\left(\frac{dC}{dx}\right)^2 + \frac{\kappa_1^{\alpha}}{2}\left(\frac{d\eta_1^{\alpha}}{dx}\right)^2 + \frac{\kappa_2}{2}\left(\frac{d\eta_2^{\alpha}}{dx}\right)^2 \right] dx \qquad (3)$$

where $\Delta f\left(\eta_1^{\alpha}, \eta_2^{\alpha}, C\right)$ is the free energy density difference at boundaries and within the grains. The definition of grain boundary energy (3) automatically takes into account of solute segregation to grain boundaries.

Similarly, the interphase boundary energy between an α-grain with orientation 1 and a β-grain with orientation 1 is given by

$$\sigma_{int}^{\alpha\beta} = \int_{-\infty}^{+\infty} \left[\Delta f\left(\eta_1^{\alpha}, \eta_1^{\beta}, C\right) + \frac{\kappa_C}{2}\left(\frac{dC}{dx}\right)^2 + \frac{\kappa_1^{\alpha}}{2}\left(\frac{d\eta_1^{\alpha}}{dx}\right)^2 + \frac{\kappa_1^{\beta}}{2}\left(\frac{d\eta_1^{\beta}}{dx}\right)^2 \right] dx \qquad (4)$$

(2) The Kinetic Equations

By defining orientation and composition field variables, the kinetics of microstructural evolution can be described by their spatial and temporal evolution. In the present model, the evolution kinetics of these field variables are described by the generalized time-dependent Ginzburg-Landau (TDGL) equations [5] which are the familiar continuum Allen-Cahn [5] and Cahn-Hilliard [4] equations,

$$\frac{d\eta_i(r,t)}{dt} = -L_i \frac{\delta F}{\delta \eta_i(r,t)}, \qquad i = 1,\ 2,\ \ldots,\ p \qquad (5a)$$

$$\frac{dC(r,t)}{dt} = \nabla\left\{ D\nabla\left[\frac{\delta F}{\delta C(r,t)}\right]\right\} \qquad (5b)$$

where L_i and D are kinetic coefficients related to grain boundary mobility and atomic diffusion coefficient, t is time, and F is the total free energy given in equation (2).

(3) Construction of the Local Free Energy Density

In order to solve the kinetic equations and thus simulate the microstructural evolution for a given system, the thermodynamics of a system has to be specified. For this purpose, the free energy functional, f_o, needs to be constructed. An example of local free energy density f_o has a form:

$$f\left(\eta_i, C(r)\right) = -\frac{A}{2}[C(r)-C_m]^2 + \frac{B}{4}[C(r)-C_m]^4 + \frac{D_\alpha}{4}[C(r)-C_\alpha]^4$$
$$+\frac{D_\beta}{4}\left[C(r)-C_\beta\right]^4 + \sum_{i=1}^{p}\left\{-\frac{\gamma_\alpha}{2}\left[C(r)-C_\beta\right]^2\left[\eta_i^\alpha(r)\right]^2 + \frac{\delta_\alpha}{4}\left[\eta_i^\alpha(r)\right]^4\right\}$$
$$+\sum_{i=1}^{q}\left\{-\frac{\gamma_\beta}{2}[C(r)-C_\alpha]^2\left[\eta_i^\beta(r)\right]^2 + \frac{\delta_\beta}{4}\left[\eta_i^\beta(r)\right]^4\right\} \tag{6}$$
$$+\frac{\varepsilon}{2}\sum_\alpha\sum_\beta\sum_{i=1}^{p}\sum_{j=1}^{q}\left[\eta_i^\alpha(r)\right]^2\left[\eta_j^\beta(r)\right]^2$$

where η_i are orientation field variables, C(r) is the local composition, C_α and C_β are the equilibrium compositions of α and β phases, $C_m = (C_\alpha + C_\beta)/2$, A, B, D_α, D_β, γ_α, γ_β, δ_α, δ_β, and ε are phenomenological parameters. These parameters can be chosen to give the mutual solubilities of the two phases, the diffusivities at boundaries and in different phases, and the interphase and grain boundary energies.

COMPUTER SIMULATION EXAMPLES

To solve the kinetic equations, equations (5) were discretized in the space and t and numerically solved in two dimensions by explicit method. For starting a computer simulation, one may either input a predefined initial two-phase microstructure or generate the initial condition by assigning small random values to all the orientation field variable and the overall average composition to the composition variable at all positions. Because our main interest is the late stage coarsening kinetics, the initial condition will not effect the final results. The grid size of simulation is 512x512 in two dimension.

In the computer simulation, the following parameter values are assumed: $C_\alpha = 0.05$, $C_\beta = 0.95$, $L_i = 1.0$ (isotropic grain boundary mobility), $\Delta x = 2.0$, $\Delta t = 0.09$. Diffusivity D can be different in the two phases for different applications.

Figure 1 is the temporal microstructure evolution in a solid-solid mixture. In this simulation, the volume fraction of α-phase (black phase) is 25% and five orientation variables were assigned to α-phase and ten to β-phase respectively. The constants were chosen as: A=2.0, B=9.8, $D_\alpha = D_\beta = 0.25$, $\gamma_\alpha = \gamma_\beta = 1.0$ $\delta_\alpha = \delta_\beta = 1.0$ and $\varepsilon = 1.0$; $\kappa_C = 0.5$, $\kappa_i^\alpha = \kappa_i^\beta = 0.5$, and D=0.5, which was assumed to be same for two solid phases. In this system, the ratio of interface energy with grain boundary energy is $\sigma_{gb} / \sigma_{in} = 0.9$, hence, the two phases do not wet each other.

Figure 2 shows the simulation microstructures in a solid-liquid mixture with 25, 50, 70 and 90% solid respectively. In a solid-liquid mixture, the liquid phase does not need any orientation variable to define it, since there is no orientation difference within the liquid. Hence, only ten orientation variables were assigned to solid phase. In this simulation, constants were chosen as: A=2.0, B=9.8, $D_\alpha = D_\beta = 0.25$, $\gamma_\alpha = \gamma_\beta = 2.0$ $\delta_\alpha = \delta_\beta = 1.0$ and $\varepsilon = 3.0$; $\kappa_C = 2.0$, $\kappa_i^\alpha = \kappa_i^\beta = 2.0$, which give $\sigma_{gb} / \sigma_{in} = 2.2$, as a result, the solid phase is totally wetted by the liquid phase. The diffusivity D is different in solid and liquid and the ratio of diffusivity in liquid to that in the solid was chosen to be 10^4. It can be seen that the particle shape automatically accommodates each other as the volume fraction of solid increases, and hence, the Ostwald ripening kinetics can be studied by this kinetic model in the whole volume fraction regime. An example of coarsening kinetics in a 25% solid system is shown in Fig. 3. It is shown that the kinetics follows the LSW [6,7] theory in the scaling regime for this volume fraction and the exponent n is 3. In the high

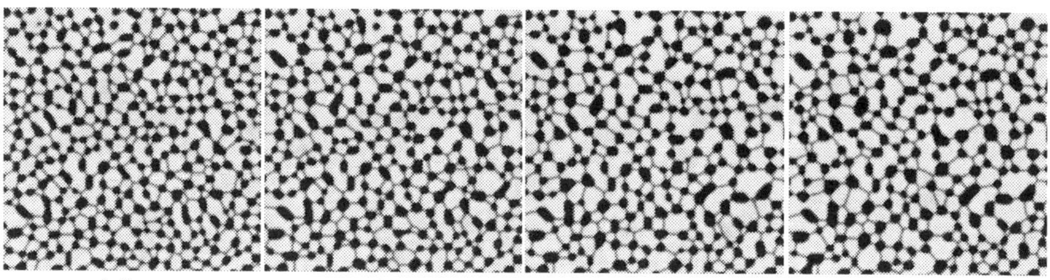

(a) (b) (c) (d)

Fig. 1. The temporal microstructure evolution in a two-solid mixture. (a) t = 200, (b) t = 500, (c) t = 800, (d) t = 1000,

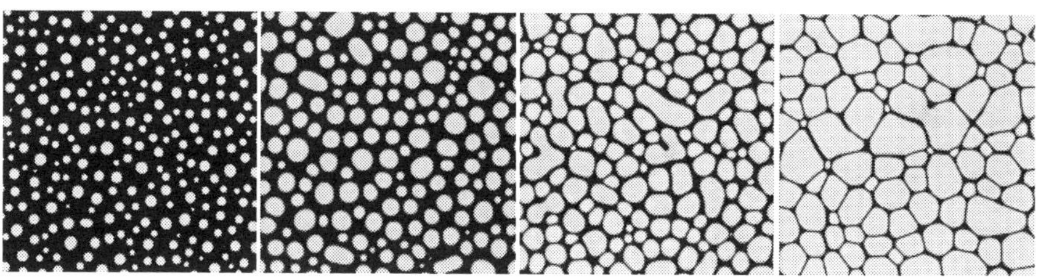

(a) (b) (c) (d)

Fig. 2. The simulated microstructure in a solid-liquid mixture. (a) 25% solid, (b) 50% solid, (c) 75% solid, (d) 90% solid.

volume fraction regime, the exponent n may change[8,9]. The detailed study will be reported in the future publications.

CONCLUSIONS

A diffuse interface model based on the continuum Ginzburg-Landau model is developed for investigating the kinetics of microstructural evolution in multiphase systems. All the statistical information about a microstructure such as grain size, grain size distribution and topological changes may be obtained from the simulated microstructures. Our preliminary results on the kinetics of Ostwald ripening in a two-phase solid-liquid mixture show that the model is very promising.

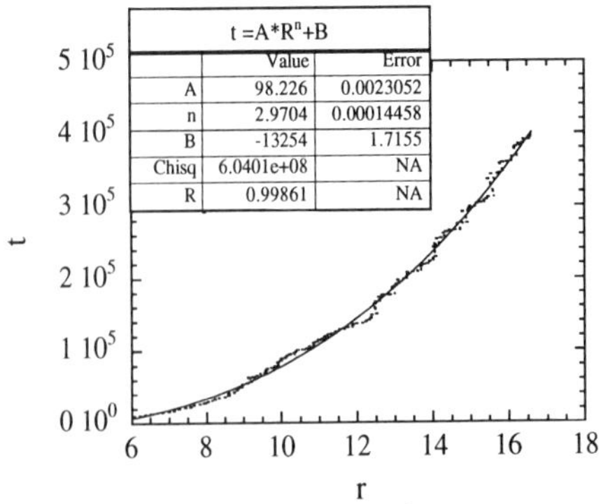

Fig. 3. The relation of average radias r with t in 25% solid mixture. The dot line is measured data, the solid line is the curve fitted with equation $t = A(r^n - r_o^n)$ and n=3.

ACKNOWLEDGMENTS

This work is supported by NSF under the grant number DMR 93-11898.

REFERENCES

1. L. Q. Chen and W. Yang, "Computer Simulation of the Domain Dynamics of a Quenched System with a Large Number of Nonconserved Order Parameters: Grain Growth Kinetics", Phys. Rev. B, **50**,15752-56 (1994).
2. L. Q. Chen, "A Novel Computer Simulation Technique for Modeling Grain Growth", Scripta metall. et mater., **32**, 115-120 (1995).
3. L. Q. Chen,"A Computer Simulation Model for Coupled Grain Growth and Ostwald Ripening in Multiphase Systems", submitted to J. Am. Ceram. Soc., 1994.
4. J. W. Cahn and J. E. Hilliard, "Free Energy of a Nonuniform System. I. Interfacial Free Energy", J. Chem. Phys. **28**, 258 (1958).
5. S. M. Allen and J. W. Cahn, "A Microscopic Theory for Antiphase Domain Boundary Motion and Its Application to Antiphase Domain Coarsening", Acta Met. **27**, 1085- (1979).
6. I. M. Lifshitz, and V. V. Slyozov, J. Phys. Chem. Solids, **19**, 35(1961).
7. C. Wagner, Z. Elektrochemie, **65**, 581 (1961).
8. P. W. Voorhees, "Ostwald Ripening of Two-Phase Mixtures", Annu. Rev. Mater. Sci., **22**, 197-215 (1992), and references therein.
9. C. K. L. Davies, P. Nash, and R. N. Stevens, "The effect of volume fraction of precipitate on Ostwald ripening," Acta Met., **28**, 179(1980).

NUMERICAL SIMULATION OF GRAIN SIZE DISTRIBUTIONS IN LIQUID PHASE SINTERED MATERIALS

Veena Tikare
Sandia National Labs., Albuquerque, NM 87185-1411
and
James D. Cawley
Case Western Reserve Univ., Cleveland, OH 44106

ABSTRACT

Grain growth in liquid phase sintered materials by Ostwald ripening has been modeled extensively by both analytical and numerical techniques. However, all models make simplifying approximations to make the problem tractable and the approximations used in these models make them most accurate at very low solid fractions. A two-dimensional, Monte Carlo simulation technique based on the Potts model that makes no assumptions about solid fraction, grain shapes or diffusion fields around grains has been used to study grain growth in fully wetting, liquid phase sintered systems. The grain size distribution, GSD, was found to vary with solid fraction, becoming broader and more peaked with increasing solid fraction. The skewness was near zero at solid fraction of 0.41 and shifted to larger grain sizes with increasing solid fraction.

INTRODUCTION

Many technologically important ceramics such as silicon nitride structural components, alumina substrates and barium titanate electrical capacitors are liquid phase sintered. It is important to understand evolution of microstructural features generated by processes such as grain growth so that these materials may be engineered for their respective applications. Greenwood[1] was the first to consider grain growth by Ostwald ripening, but the approach of Lifshitz & Slyozov[2] and of Wagner[3] (LSW theory) is the basis of modern studies of this subject. In order to make the model tractable, the LSW theory assumes the distance between grains is much larger than the grain size (or equivalently the fraction of grains is zero) and a mean solute concentration in the liquid matrix. The LSW theory predicts a power law behavior for grain growth given by the following:

$$R^n - R_o^n = kt \qquad \textbf{eq. 1}$$

where t is time, R is grain radius at time t, R_o is the grain radius at t = 0, k is a constant and, the grain growth exponent n = 3 for diffusion controlled grain growth. In addition, LSW predicts all systems will evolve to a unique steady state grain size distribution, GSD. Many investigators[4,5,6,7] have modified the LSW theory to include finite volume fraction of grains using a variety of techniques; however, they all make simplifying assumptions such as spherical grains at all solid fractions[5-8], uniformly spaced grains[5,6], mean field concentration[5,6,8], and all grains of a given size experience the same local environment[5,6]. All have found the power law behavior given is eq. (1) to hold with the same exponent, n = 3, as well as broadening and increasing skewness of the GSD with increasing solid fraction. However, the GSD predicted by each of these models varies. We attribute these variations to the various assumptions made by each. In this paper, the GSD's obtained by a Monte Carlo approach are presented and compared to those predicted by other grain growth theories.

SIMULATION METHOD

The Q-state, Potts model adapted by Srolovitz et al.[8] to study grain growth was modified to simulate grain growth in liquid phase sintered materials. The microstructure was represented digitally on a two-dimensional, square lattice with periodic boundary conditions in both the X- and Y- directions. Two phases, liquid matrix and solid grains, were simulated by populating the lattice with a two-component canonical ensemble with the two components designated as A and B. Component A, the primary component in the solid phase, was allowed to assume one of 100 degenerate states; while component B, the primary component in the liquid, was allowed to assume only one state. These states may be considered crystallographic orientation for the purposes of these simulations. The Hamiltonian, H, for the simulation was the sum of the interaction energy, J, between each site, i, and each of its eight first and second nearest neighbors, j, as given by

$$H = \frac{1}{2} \sum_{j=1}^{8} \sum_{i=1}^{N} J(q_i, q_j) \qquad \textbf{eq. 2}$$

where q_i is the orientation of site i and N is the total number of sites. The interaction energies used in these simulations were J(B,B) = 0, J(A,B) = 1.0, $J(A_i, A_i) = 0$ and $J(A_i, A_j) = 2.5$. These interaction energies where chosen to give full wetting in the simulation system. The simulation steps were (1) choose a site at random; (2) choose a neighboring site at random; (3) if one is an A-site and the other a B-site, choose a new orientation for the A-site at random; if not, go to step 6; (4) consider the two sites exchanging places with each other; (5) use the Metropolis algorithm to determine the exchange probability, P;

$$P = \exp - \frac{\Delta H}{k_B T} \qquad \Delta H > 0$$

$$P = 1 \qquad \Delta H \leq 0 \qquad \textbf{eq. 3}$$

where ΔH is the change in energy given by eq. (2), k_B is Boltzmann's constant and T is absolute temperature; (6) increment the number of steps by 1/N; return to step (1) if the number of iterations is less than that desired. This simulation method resulted in material transport from one grain to another by solution at the receding grain, diffusion through the liquid matrix, and precipitation onto the growing grain. No solid state diffusion occurred.

The simulations were run at temperature, $k_B T = 1.3$. The solid fraction, Ø, was varied from Ø = 0.41 to 0.89 by changing the mole fraction of components A and B. Multiple simulation of size 300 X 300 sites were run at each solid fraction to generate a large number of grains.

RESULTS

Analysis of grain size as a function of time revealed the power law behavior for grain growth is obtained in these simulations with a grain growth exponent of n = 3, as predicted for diffusion-controlled grain growth. This exponent was found to be constant with solid fraction.

The GSD obtained from the simulations are plotted as normalized frequency versus normalized grain size, R/<R>, so that the GSD's may be directly compared. Dynamic scaling of the GSD was obtained for all solid fractions studied. The GSDs after 100,000 are used for comparison in figure 1. The GSD varied with the solid fraction. It became broader and more peaked with increasing solid fraction. The GSD was skewed to the right at solid fraction, Ø = 0.89, and the direction of skewness shifted to the left with decreasing solid fraction.

DISCUSSION

The simulation results were in agreement with grain growth theories on the trend in the spread of grain sizes. As the solid fraction increased, the GSD broadened because a few grains grew larger. The size of the largest grains

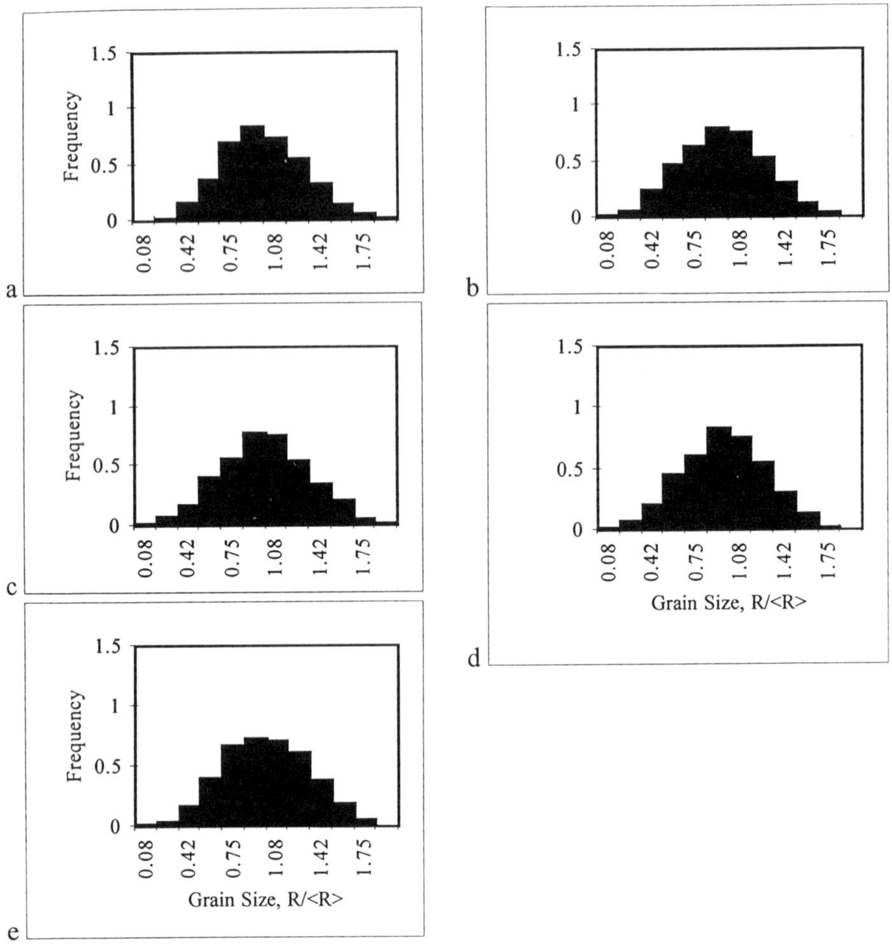

Figure 1. GSD's obtained from simulations at solid fractions of (a) 0.89, (b) 0.78, (c) 0.66, (d) 0.53 and (e) 0.41.

increased from 1.9<R> at Ø = 0.41 to 2.2<R> at Ø = 0.89. This spread was larger than GSD's predicted by all grain growth theories except for Voorhees and Glicksman's[6]

The skewness of the GSD obtained from simulation was calculated to be weakly negative, almost zero at a solid fraction, Ø = 0.41 and it increased with solid fraction to 0.3 at Ø = 0.89. This is very different from the predictions of grain growth theories. At very low solid fraction, the LSW theory predicts a GSD with a negative skewness. The theories which consider finite solid fractions predict an increase in skewness, but not a change from negative to positive skewness as observed in the simulations. Figure 2a compares the GSD from simulations at solid fraction Ø = 0.53 to that predicted by Voorhees and Glicksman. While, the spread in grain sizes were in agreement, the skewness was quiet different. The GSD obtained from simulations was skewed slightly to the larger sizes while Voorhees and Glicksman's was skewed slightly to smaller sizes. The simulation GSD at Ø = 0.66 is compared to experimental data collected by Fang and Patterson[9] for the W-14Ni-6Fe system at Ø = 0.60 in figure 2b. They allowed the GSD to reach the steady state GSD and measured grain size by image analysis of two-dimensional cross-sections. Although, the simulation GSD was in good agreement with the experimental data; the simulation predicts a more peaked GSD. This difference is ascribed to the difference in dimensionality. Prior work has shown the two-dimensional models routinely give more peaked GSD's than three-dimensional ones[10,4,7].

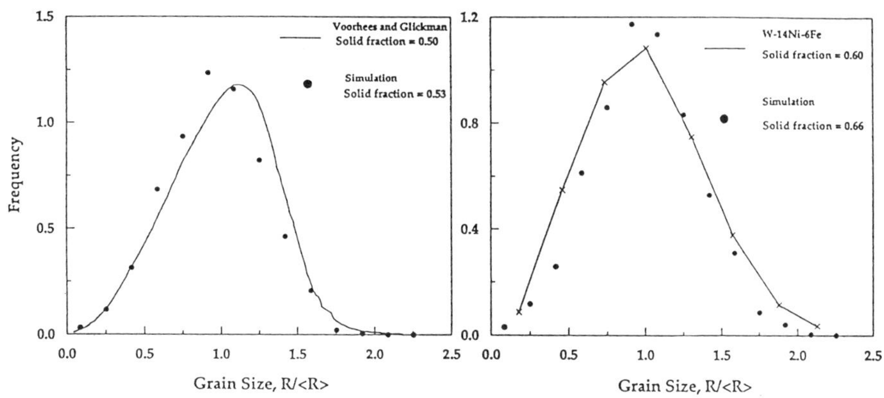

(a) (b)

Figure 2. Comparison of GSD obtained from simulations to that obtained (a) by Voorhees and Glicksman and (b) experimentally by Fang and Patterson.

The advantage of the simulation technique presented in this investigation over that of other models was the spatial distribution of grains was allowed to develop naturally. Further, grain shape accommodation and solute gradients between grains were inherent to the technique. Voorhees & Glicksman[6] and Yao et al.[7] considered spatial arrangement of grains, but they only considered diffusion gradients around spherical grains, at relatively low grain fractions where interactions were minimal. DeHoff's[11] communicating neighbor model is, in principle, able to consider both deviation in curvature of grains and interactions of gradients between grains, but correlating curvature parameters and interaction parameters to nonspherical grain shapes has not been done. The ability of the simulation to incorporate spatial features may be the reason it was able to predict the change in skewness of the GSD from negative to positive with increasing solid fraction.

CONCLUSIONS

Grain Size distributions were characterized over a range of solid fractions, 0.40 to 0.90. Under isotropic, fully wetting conditions, the grain size distribution varied with the solid fraction. It became broader and more peaked with increasing solid fraction. The grain size distribution was skewed to larger sizes at solid fraction = 0.89. As the solid fraction decreased, the skewness shifted toward smaller sized grains.

ACKNOWLEDGMENT

This work was performed primarily under a NASA/CWRU cooperative agreement on ceramics processing (NCC3-139) and completed by one of us (V.T.) at Sandia National Labs. while supported by the U.S. Dept. of Energy under contract number DE-AC04-94AL85000.

REFERENCES

[1]G.W. Greenwood, "The Growth of Dispersed Precipitates in Solution," Acta Metall., 4 243-248 (1956).
[2]I.M. Lifshitz and V.V. Slyozov, "The Kinetics of Precipitation From Supersaturated Solid Solutions," J. Phys. Chem. Solids, 19 [1-2] 35-50 (1961).
[3]C. Wagner, "Theory of the Aging of Precipitates by Redissolution (Ostwald Ripening)," Zeitschrift fur Elektrochemie, 65 [7-8] 581-591 (1961).
[4]A.J. Ardell, "The Effect of Volume Fraction on Particle Coarsening: Theoretical Considerations," Acta Metall., 20 61-71 (1972).
[5]A.D. Brailsford and P. Wynblatt, "Dependence of Ostwald Ripening Kinetics on Particle Volume Fraction," Acta Metall., 27 489-497 (1979).
[6]P.W. Voorhees and M.E. Glicksman, "Solution to the Multi-Particle Diffusion Problem with Application to Ostwald Ripening - II. Computer Simulations," Acta Metall., 32 [11] 2013-2030 (1984).
[7]J.H. Yao, K.R. Elder, H. Guo, and M. Grant, "Theory and Simulation of Ostwald Ripening," Phys. Rev., B 47 [21] 14110-14125 (1993).
[8]D.J. Srolovitz, M.P. Anderson, G.S. Grest and P.S. Sahni, "Grain Growth in Two Dimensions," Scripta Metall., 17 241-246 (1983).
[9]Z. Fang and B.R. Patterson, "Experimental Investigation of Particle Size Distribution Influence on Diffusion Controlled Coarsening," Acta Metall., 41 [7] 2017-2024 (1993).
[10] A.J. Ardell, "Isotropic Fiber Coarsening in Uniderectionally Solidified Eutectic Alloys," Met. Trans., 3 [6] 1395-1401 (1972).
[11]R.T. DeHoff, "A Geometrically General Theory of Diffuson Controlled Coarsening," Acta Metall., 39 [10] 2349-2360 (1991).

Diffusion and Mass Transport

ATOMISTIC MOBILITY IN SUPERCOOLED LIQUIDS VIA NORMAL MODE ANALYSIS AND MOLECULAR DYNAMIC SIMULATIONS

Brian J. Reardon, John Kieffer, University of Illinois, Dept. of Materials Science and Engineering, 105 S. Goodwin Ave., Urbana, IL 61820; breardon@uiuc.edu, johnk@uiuc.edu

ABSTRACT

Normal mode analysis carried out on disordered structures, such as of a liquid, reveals a distribution of imaginary frequencies. Using the NaCl melts as a model, the nature of these imaginary frequencies was investigated by identifying the corresponding time correlation functions in dynamic simulations. Accordingly, imaginary frequencies represent the rates of divergence in phase space trajectories, which result from slight perturbances. The largest imaginary frequency dominates the loss of correlation in the time evolution of these systems, and is key to the evaluation of transport coefficients.

INTRODUCTION

The significant increase in the ensemble size for molecular scale simulations, which is possible with today's computers, allows for a number of new approaches to materials research. In the field of ionic conductors, great emphasis is put upon understanding the relationship between the mobility of the charge carrying species and the geometry of the embedding structure. Since ionic conduction takes place within a structure in thermal equilibrium, it can be expected that transport properties depend on the force fields created by all atoms in the structure at their average positions. Recently, there has been an upsurge of investigations where normal mode analysis (NMA) was carried out on thermally disordered structures, promising to provide conclusive insight into this question.[1-9]

BACKGROUND

Statistical mechanics provides a number of formalisms for extracting transport coefficients form the phase space trajectory of a dynamical system. All of these methods are based in one way or another on the evaluation of time correlation functions. For example, the diffusion coefficient can be obtained by integration of the velocity autocorrelation function,

$$D = \frac{1}{3}\int_0^\infty \langle \mathbf{v}(t)\mathbf{v}(0)\rangle dt \tag{1}$$

Alternatively, one can use the mean squared displacement, which represents the correlation in the positions of particles,

$$D = \frac{1}{6t}\left\langle |\mathbf{r}(t) - \mathbf{r}(0)|^2 \right\rangle \tag{2}$$

Another correlation function which also yields the diffusion coefficient is the self-component of the intermediate scattering function,

$$F_s(\mathbf{q},t) = \frac{1}{N}\left\langle \delta_{jk} \sum_{j=1}^{N} e^{i\mathbf{q}\mathbf{r}_j(t)} \sum_{k=1}^{N} e^{i\mathbf{q}\mathbf{r}_k(0)} \right\rangle \equiv e^{-q^2 Dt}, \tag{3}$$

which describes the correlations between the Fourier components of the density operator, and whose experimental counterpart is the incoherent neutron scattering intensity. It can be shown that provided the motion is random, all these methods are equivalent.

Obviously, by their mere definition, these methods require one to observe a simulation ensemble over an extended period of time. An interesting question, however, arises for ergodic systems, i.e., systems large enough that any dynamical variable exhibits a distribution of values, and this distribution is invariant in time. In this case, a single configuration contains all the information that a long-time simulation could reveal. This condition should be more closely fulfilled, the larger the ensemble, and one may anticipate that the system sizes which are currently possible, provide acceptable accuracy. In that case, one could explore the frequency spectrum of the normal modes of motion to estimate the magnitudes of transport coefficients, based on the following argument.

NMA is based on solving the classical equations of motion for a configuration of N atoms,

$$m_j \ddot{\mathbf{u}}_j = -\sum_{k=1}^{N} \nabla\phi(\mathbf{r}_{jk}) = -\sum_{k=1}^{N} \nabla\phi(\mathbf{r}_{jk}) - \frac{1}{2}\sum_{k=1}^{N} \nabla^2\phi(\mathbf{r}_{jk})\mathbf{u}_k - O(u^3), \tag{4}$$

where $\mathbf{r}_{jk} = \mathbf{r}_j - \mathbf{r}_{jk}$ and $\mathbf{r}_j$ is the position of particle j, and $\mathbf{u}_j$ represents a small displacement from that position. If the particle positions correspond to those of an equilibrium configuration, the first term on the right hand side vanishes, and the solution for the displacements as a function of time is given by

$$\mathbf{u}_{j,\nu} = \boldsymbol{\varepsilon}_{j,\nu} e^{i(\mathbf{q}\mathbf{r}_j - \omega_\nu t)}, \tag{5}$$

where the index ν represents a particular mode of motion. The frequency corresponding to this mode is obtained by solving the eigenvalue problem

$$\Delta\boldsymbol{\varepsilon} = \omega^2 \boldsymbol{\varepsilon}, \tag{6}$$

which arises after substituting Eq.(5) into Eq.(4), and where Δ is a matrix whose elements are the partial derivatives $\Delta_{jk} = \frac{\partial^2 \phi}{\partial_\alpha u_j \partial_\beta u_k} e^{i\mathbf{q}\mathbf{r}_{jk}}$. Finally, the actual motion of any given particle is given by the linear combination of the displacements corresponding to all modes,

$$\mathbf{u}_j(t) = \frac{1}{\sqrt{Nm_j}} \sum_{\nu=1}^{N} c_{j,\nu} \mathbf{u}_{j,\nu} = \frac{1}{\sqrt{Nm_j}} \sum_{\mathbf{q}} \sum_{\nu=1}^{N} c_{j,\nu} \boldsymbol{\varepsilon}_{j,\nu} e^{i(\mathbf{q}\mathbf{r}_j - \omega_\nu t)}, \tag{7}$$

where $c_{j,\nu}$ are coefficients which describe the contribution of the νth mode to the amplitude of motion of particle j, and which are subject to the constraint imposed by the total thermal energy stored in the system,[10]

$$\left\langle \left| c_{j,\nu}(\mathbf{q}) \right|^2 \right\rangle = k_B T / \omega_\nu^2(\mathbf{q}). \tag{8}$$

Taking the derivative of Eq. (7) with respect to time yields the velocity with which particle j moves about its equilibrium position. Using this velocity to evaluate the velocity autocorrelation function yields

$$\left\langle \dot{\mathbf{u}}_j(t) \dot{\mathbf{u}}_j(0) \right\rangle = \frac{1}{Nm_j} \sum_{\mathbf{q}} \sum_{\nu=1}^{N} \omega_\nu^2(\mathbf{q}) \left\langle c_{j,\nu}(\mathbf{q},t) c_{j,\nu}(-\mathbf{q},0) \right\rangle e^{i\omega_\nu t} = \frac{k_B T}{Nm_j} \sum_{\mathbf{q}} \sum_{\nu=1}^{N} e^{i\omega_\nu t}. \tag{9}$$

Since the summation on the right hand side goes over all possible frequencies the system is capable of, this equation is the basis for the widely used assumption that the Fourier spectrum of the mass weighted velocity autocorrelation function is equal to the phonon density of states (DOS). Provided that this assumption is accurate, then the inverse holds and one can derive the velocity correlation function from the DOS, and with Eq.(1), the diffusion coefficient.

There are, however, several simplifications in linking Eqs. (1) and (9), which one has to evaluate with caution. First of all, it was assumed that the velocity of every particle at time t could be evaluated from its velocity at time 0, based on a simple phase factor ωt. This would imply that all motion was purely harmonic, and ultimately it would mean that no diffusion could occur. Note that Eq.(1) can be viewed as the Fourier spectrum of vibrational modes at zero frequency. Except for three translational modes, which are not diffusive, the vibrational DOS at $\omega = 0$ vanishes. A non-zero intercept of the ordinate in the DOS is the result of damping in the velocity correlation function. The physical origin of such damping are phonon collisions where energy is transferred between vibrational modes. This process is based on the anharmonic components of the interaction potential. Hence, in the conventional approach to NMA, which is carried out in the harmonic approximation, using the atoms of the primitive cell at their equilibrium position, anharmonic effects cannot be encompassed.

If on the other hand, NMA is carried out on a large ensemble, which is thermally disordered so that no two primitive cells are exactly the same, a more extended portion of the potential energy minimum is mapped out in the process of evaluating the force constants. In this case one might expect that the dynamical matrix, in some form, contains information about the variation in the force constants with variable interatomic spacings, and hence, reflect the effects of anharmonicity in the interaction potential.

RESULTS AND DISCUSSION

Fig. 1 shows the results of this type of analysis, labeled by its pioneers as instantaneous normal mode analysis (INMA).[1-9] In accordance with the approach made in Eq. (5), NMA yields positive eigenvalues for the squares of vibrational frequencies. However, INMA of disordered structures also yields negative eigenvalues, whose square roots, when substituted in Eq. (5), represent the rates of non-periodic motion. (In this paper we will use the terms imaginary frequencies and divergence rates equivalently.) In Fig. 1, the distribution of vibrational frequencies are plotted on the positive axis, and that of the roots of negative eigenvalues are plotted on the negative axis. Note that this representation ignores the fact that both, positive and negative roots of all rates of motion are solutions of the equations of motion. This is irrelevant in the case of oscillatory motion, but in the case of non-periodic motion the sign of the rate of motion differentiates between vastly different temporal evolutions.

Based on the occurrence of negative eigenvalues, the originators of the INMA have overcome the difficulties described in the previous section, and have developed methodologies for deriving the self diffusion coefficient from the DOS.[1-9] Their approach begins with the equation proposed by Zwanzig,[11]

$$D = \frac{k_B T}{3Nm_j} \int_0^\infty \int_0^\infty \rho(\omega) \cos(\omega t) e^{-\gamma t} d\omega dt \qquad (10)$$

which can be understood as a simple correction of the time correlation function which would result from perfectly harmonic motion, $\langle \mathbf{v}(t)\mathbf{v}(0) \rangle = \langle |\mathbf{v}|^2 \rangle \cos(\omega t)$, by means of an exponential factor containing an effective hopping rate γ. Keyes, Stratt, Laird, and their respective co-workers obtain the value for this hopping rate from the lobe of imaginary frequencies, using the average value, the relative fraction of non-periodic modes, and a Boltzmann factor containing the activation energy for hopping.[1-9] While the model described by these authors is intuitively very attractive, in this study we have taken a somewhat different approach towards the interpretation of imaginary frequencies resulting from INMA. In the following, we would like to offer an analysis which delineates the basis for our thinking.

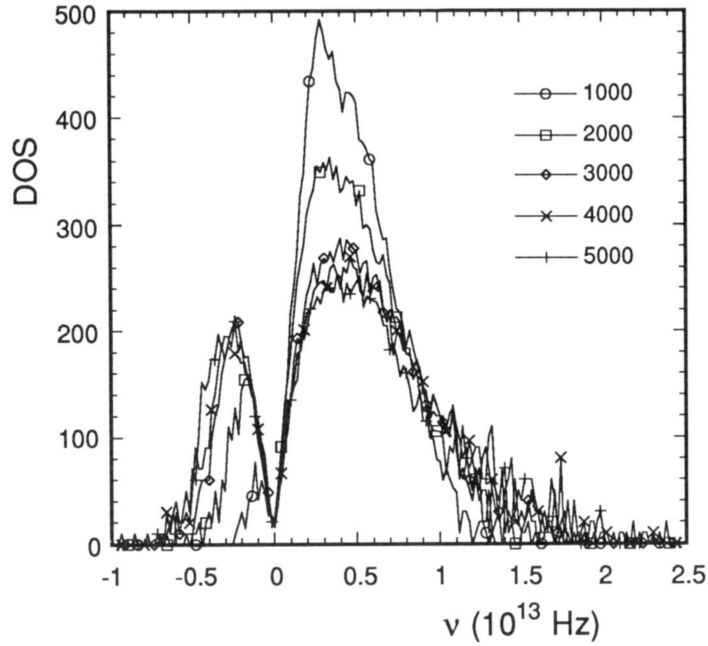

Fig. 1 Distribution of normal mode frequencies for NaCl at various temperatures. Positive values correspond to oscillatory frequencies, negative values to convergent or divergent trajectories.

Recall that NMA reveals the directions and rates of motion of a system which is minutely displaced from its equilibrium, or in more general terms, the temporal evolution of the perturbance which displaces a dynamical system from a given phase space trajectory. Negative rates of non-periodic motion therefore reflect an exponential return to the original trajectory, which implies that correlations in the phase space trajectory are restored as a result of these modes. Positive rates, on the other hand, indicate an exponential departure from the original phase space trajectory, and consequently, destruction of correlation in that motion. This is schematically illustrated in Fig. 2.

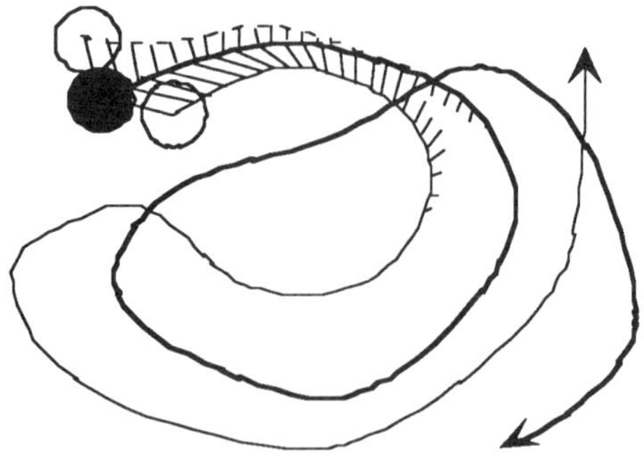

Fig. 2 Schematic of perturbed atomic trajectories. Let the thick line be the original trajectory of a particle caged by its neighbors (which are not shown). The thinner lines represent trajectories if the particle was slightly displaced at the start. The dashed thin line shows the effect of a negative normal mode rate, and the solid thin line that of a positive rate. In the first case, the atom is soon back in phase with the original motion, whereas in the second case, the positive rate of deviation causes complete separation from the original trajectory, and dephasing of the motion.

The occurrence of such divergent trajectories can be demonstrated using molecular dynamic simulations, by computing the appropriate correlation functions. First, the velocities of all particles are recorded for a typical trajectory. Then the procedure is repeated after all particles have been minutely displaced in the starting configuration. The mean squared displacement from the original positions can be equated to a temperature fluctuation using Eq. (8). The desired correlation function is obtained by forming the dot product between the velocity vectors of each particle, along the original (**v**(t)) and the perturbed trajectories (**v**'(t)), and averaged over all atoms in the system, i.e., $\langle \mathbf{v}(t)\mathbf{v}'(t)\rangle_{ens.}$. This correlation function reveals how fast, on average, atoms deviate from a given path if they are disturbed in their motion. Perturbances can simply be attributed to temperature fluctuations. Results are shown in Fig. 3.

It is interesting to observe that the loss of correlation between the velocities corresponding to the perturbed and original trajectories is not exponential. The smaller the temperature fluctuation, the longer the incubation period during which the perturbed motion is in phase with the unperturbed one. Eventually, however, correlation disappears, which is to be expected for non-linear dynamical systems. The rate of the drop in these correlation functions seems to increase slightly when using very small disturbances (the logarithmic scale in Fig. 3 over-emphasizes this). Even for temperature fluctuations as small as 0.05 K, the loss of correlation is complete at about 1 ps. This duration coincides with the decay of the dashed curve in Fig. 3, which is an exponential based on the largest divergence rate, γ_m, in the DOS spectrum. Moreover, the slowest decay is exhibited by the curve corresponding to $\int_0^\infty \rho(\gamma)e^{-\gamma t}d\gamma$.

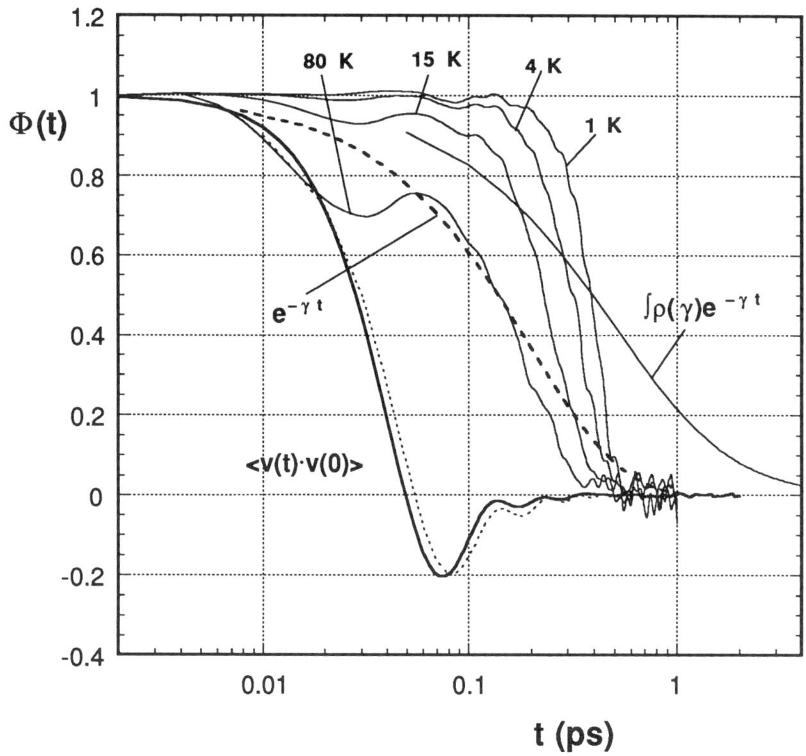

Fig. 3 Various time correlation functions determined from a simulated NaCl configuration at 2000 K. The curves labeled with $\langle \mathbf{v}(t)\mathbf{v}(0)\rangle$ are the velocity correlation function for Na (solid line) and the one determined from the DOS (dotted line). The dashed curve represents the function used to modulate the DOS. The curves labeled with temperatures represent the correlations between perturbed and unperturbed velocities, where each temperature reflects the amplitude of the imposed fluctuation at time 0. The solid curve to the right represents the relaxation based on the entire spectrum of imaginary frequencies.

The dotted curve represents a velocity correlation function calculated using Eq. (10), where the room-temperature DOS of a quenched NaCl melt was used for $\rho(\omega)$, and γ_m for γ. The curve generated in this way is in close agreement with the actual velocity autocorrelation function of Na derived from the MD simulation. (The correlation function for Cl is quite similar, only slightly shifted towards the right.) Using the positive lobe of the DOS at 2000 K produced rather unsatisfactory results, due to the high frequencies present in this distribution. Furthermore, using the weighted effect of all divergence rates results in phonon damping about an order of magnitude slower than actually exhibited by the velocity autocorrelation function.

CONCLUSION

We have shown that the velocity autocorrelation function of the components of liquid NaCl can be reconstructed based on the fundamental vibrational frequencies of the structure near its potential energy minimum, and the largest divergence rate exhibited by the negative eigenvalue lobe resulting from INMA at the actual temperature. This finding indicates that the rate of loss of correlation in atomic motion is controlled by the largest divergence rate, as it reflects the strongest instability in the dynamical system and acts as the 'weakest link' in maintaining a given trajectory. The contribution to phonon damping by a given diverging mode is not commensurate to the weight reflected by the negative lobe of the DOS.

REFERENCES

1 P. Moore, T. Keyes, J. Chem. Phys., 100, 9, ,6709, (1994).

2 G. Seeley, T. Keyes, J. Chem. Phys., 91, 9, 5581, (1989).

3 B. Madan, T. Keyes, G. Seeley, J. Chem. Phys., 92, 12, 7565, (1990).

4 G. Seeley, T. Keyes, B. Madan, J. Chem. Phys., 95, 5, 3847, (1991).

5 G. Seeley, T. Keyes, B. Madan, J. Chem. Phys., 96, 4074, (1992).

6 T. Keyes, J. Chem. Phys., 101, 5, 5081, (1994).

7 M. Buchner, B. Ladanyl, J. Chem. Phys. 97, 11, 8522, (1992).

8 M. Cho, G.R. Fleming, S. Saito, I. Ohmine, R. Stratt, J. Chem. Phys., 100, 9, 6672, (1994).

9 S.D. Bembenek and B.J. Laird, Phys. Rev. Letters 74, 936 (1995)

10 M.T. Dove, "Introduction to Lattice Dynamics," p.61, Cambridege University Press, Cambridge, 1993

11 R. Zwanzig, J. Chem. Phys. 79, 4507 (1983)

DIFFUSION IN POLYCRYSTALLINE MICROSTRUCTURES

Thomas P. Swiler and Elizabeth A. Holm
Sandia National Laboratories, P.O. Box 5800, Albuquerque NM 87185-1411

ABSTRACT

Mass transport properties are important in polycrystalline materials used as protective films. Traditionally, such properties have been studied by examining model polycrystalline structures, such as a regular array of straight grain boundaries. However, these models do not account for a number of features of real grain ensembles, including the grain size distribution and the topological aspects of grain boundaries. In this study, a finite difference scheme is developed to study transient and steady-state mass transport through realistic two dimensional polycrystalline microstructures. The effects of microstructural parameters such as average grain size and grain boundary topology are examined, as are the effects that are due to the limits of the model.

INTRODUCTION

Grain boundary diffusion coefficients are often six orders of magnitude larger than bulk diffusivities in polycrystalline materials. Therefore, although the volume of grain boundary is a small fraction of the total, grain boundary diffusion often dominates the effective diffusivity of solutes in real materials. Such accelerated diffusion affects various applications in different ways. For example, in protective films grain boundaries act as a fast path through the material to an intolerant material beneath. In a diffusion barrier, boundaries may increase the steady state flux of a contaminant through the film. In applications in which corrosion properties are composition dependent, the penetration of species into the material along grain boundaries determines the lifetime of the specimen.

A number of analytic and numerical models for grain boundary diffusion problems have been proposed [1-10]. Most of these models use simplifying assumptions that restrict either the geometry and topology of the system or the time regime of applicability. In addition, most models ignore the effect of three-grain junctions which can act as bottlenecks to diffusive flux. To overcome these restrictions, we have developed a flexible finite difference technique that can be applied to arbitrary bitmapped microstructures across all time regimes.

METHOD

The input microstructures for our finite difference diffusion simulations are bitmapped structures where each pixel is assigned an index that identifies it as a member of a particular grain. The dual of the bitmap lattice gives a grid of lines that separate lattice sites. These gridlines and their points of intersection represent solute diffusion paths and known solute concentration during the diffusion simulation. Each line segment is assigned a diffusivity corresponding either to that of a grain boundary or that of the bulk, as shown in Figure 1. In these simulations, both the bitmap and the diffusion grid are square lattices, and grain boundaries are assumed to be one unit wide. The microstructures may have either periodic or closed boundary conditions at the edges of the lattice.

Diffusion along each path in the grid takes place according to a discretized version of the usual diffusion equation

$$\frac{\partial C}{\partial t} = D\left(\frac{\partial^2 C}{\partial x^2} + \frac{\partial^2 C}{\partial y^2}\right) \tag{1}$$

where C is the concentration, t is time, x and y are spatial coordinates, and D is the appropriate diffusion coefficient (bulk or boundary) for the path. To solve this equation, we use the Crank-Nicholson algorithm that combines implicit and explicit finite difference solutions for a total solution that is second-order accurate in the time and space discretization [11] and has good stability. We integrate in a two-dimensional lattice using the Alternating Direction Implicit (ADI) method that exchanges the implicit and explicit partial solutions between the two dimensions at each half timestep. Although this method is generally stable for long timesteps, there are two factors that limit the length of the timestep in our simulations. The first is that large transient errors may result from using long timesteps at short times in the simulation. If one is interested in the transient behavior of the system, then it is necessary to limit the time step at short times to obtain accurate solutions. The second limitation becomes apparent when modeling grain boundary diffusion. At points where the diffusion coefficient becomes discontinuous in the direction of the flux, such as at three-grain junctions, the use of very long timesteps results in spikes in the transient concentration after the first half of each timestep that cannot be accurately counteracted in the second half. Unchecked, this can result in a pile-up of solute at these points, so the length of the timestep must again be limited to reduce the error. In order to minimize the computational effort during the course of a simulation, we employ an adaptive timestep scheme that utilizes the longest possible timestep to obtain the required accuracy given both limitations.

We create finite sources or sinks by setting the initial concentration at points in the diffusion lattice as desired. We hold the concentration of these points constant to create infinite sources or sinks. The flux out of or into infinite sources or sinks can be measured so that the steady state flux in various diffusion problems can be determined.

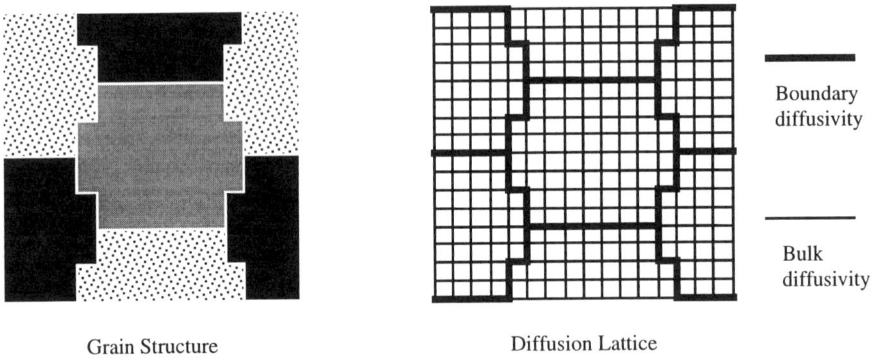

Figure 1. The method by which a bitmapped microstructure is mapped to a diffusion grid. Note that diffusivities along grain boundaries correspond to the grain boundary coefficient, diffusivities across grain boundaries correspond to the bulk coefficient.

VALIDATION

We validated our simulation technique by comparing to a number of analytic results. We found excellent agreement across all time regimes for diffusion from an infinite source into a finite single crystal, from a semi-infinite source into a semi-infinite crystal, from arrays of thick film solute sources, and others [11].

Because the grain boundaries in our system are a single unit wide, the grain boundary cannot support a concentration gradient. To determine if this discretization affects diffusion results, we compared simulation results to the analytic solution for diffusion in a single crystal with a periodic array of thin film solute sources [11]. In the simulation, the solute sources were grain boundaries a single unit wide with the same diffusion coefficient as the matrix but with different initial solute concentration from the matrix. As shown in Figure 2(a), the simulation shows excellent agreement with the exact solution.

Another concern is stability and accuracy of the simulation when diffusivities have discontinuities. To examine this case, we compared simulation results to the analytic solution for a semi-infinite slab solute source with diffusivity D_1 bonded to another semi-infinite slab with diffusivity D_2 [11]. As shown in figure 2(b), the simulation is in excellent agreement with the exact solution, even near the diffusivity discontinuity located at $x = 500.5$.

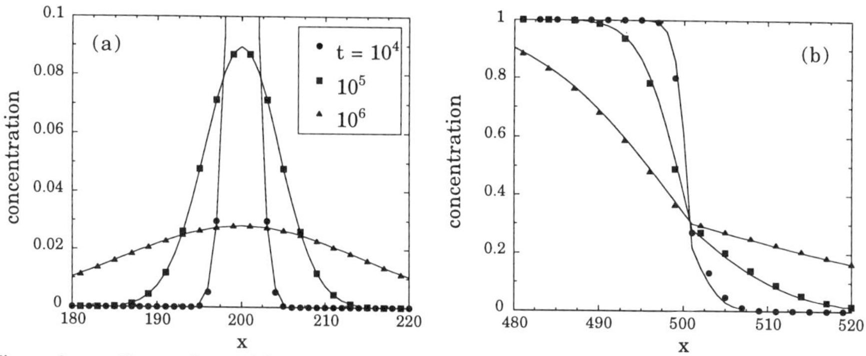

Figure 2. Comparison of finite difference calculation results (lines) to analytic results (points) for (a) diffusion from a thin-film source at $x = 200$ and (b) diffusion in a heterogeneous diffusion couple with an interface at $x = 500.5$.

STEADY STATE DIFFUSION

We utilized the validated finite difference technique to study steady state diffusive flux through polycrystalline microstructures. Realistic two-dimensional grain structures were generated by Monte Carlo Potts models simulations for grain growth on the square lattice with first and second neighbor interactions [12, 13]. In this case, we wish to compare the effective diffusion coefficient for a realistic microstructure to that for an idealized system.

Consider a system with an infinite solute source at $x = 0$ in which all grain boundaries are parallel to the x-axis. As usual, the steady state flux of solute $J = D_{eff}\Delta c / \Delta x$, where D_{eff} is the effective diffusivity for the medium. To a first approximation, the boundary and bulk contributions to the flux may be decoupled, so D_{eff} is given by a rule of mixtures relation

$$D_{eff} = \frac{D_b \delta}{\bar{r}} + \frac{D_g}{\bar{r}}\left(1 - \frac{\delta}{\bar{r}}\right) \tag{2}$$

where D_g is the diffusion coefficient in the grain bulk, D_b is the grain boundary coefficient, δ is the grain boundary width, and $\bar{r}$ is the mean linear intercept grain size. Geometry shows that $\delta / \bar{r}$ is the area fraction of grain boundary. If $D_b\delta >> D_g$, as in this study, $D_{eff} \approx D_b\delta / 2\bar{r}$, which is plotted as the upper line in Figure 3.

Using the finite difference diffusion simulation, we computed effective diffusivities from the steady state flux through microstructures with varying average grain sizes. These data are plotted as the symbols in Figure 3. The measured effective diffusivity varies inversely with grain size as predicted by the rule-of-mixing. However, D_{eff} is smaller than the rule-of-mixing predicts at all grain sizes.

There are three factors which combine to decrease D_{eff} in the simulation microstructures. First, three-grain junctions are grain boundary diffusion flux bottlenecks. Consider, for example, a three-grain junction in which the two inlet boundaries are carrying their full flux capacities. The outlet boundary can carry only half of the total possible inlet flux. Likewise, in a junction with one inlet boundary at full carrying capacity, both outlet boundaries receive only half of the inlet flux. Monte Carlo simulations which generate a random array of both types of junctions and compute the average flux carrying capacity of the boundaries show that for large numbers of junctions, the flux carrying capacity seems to asymptote at 0.43 times the ideal capacity.

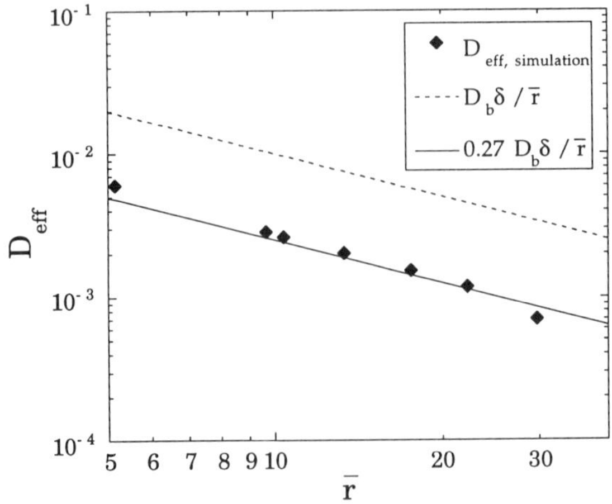

Figure 3. Effective diffusion coefficient D_{eff} as a function of grain size $\bar{r}$ for simulated microstructures with $D_b = 10^{-1}$ and $D_g = 10^{-6}$. The rule of mixing coefficient $D_{eff} = D_b\delta / \bar{r}$ and the corrected rule of mixing coefficient $D_{eff} = 0.27 D_b\delta / \bar{r}$ are shown for comparison.

In addition, two geometric factors decrease boundary flux. In an array of polygonal grains, the boundaries form an indirect path across the sample. For example, in a hexagonal array of grains, the shortest point-to-point path is $2/\sqrt{3} = 1.15$ times the distance between endpoints. For endpoints at the least favorable misorientation, the shortest path is 4/3 = 1.33 times the endpoint separation. We expect an average path to be about 1.24 times longer than the sample width, which decreases flux to about 0.8 times the ideal capacity.

The final factor is purely a simulation artifact. In the square lattice used here, each boundary is represented as a zig-zag of boundary segments parallel and perpendicular to the x-axis of the simulation lattice. Due to this discretization, in a system of randomly oriented boundaries, the boundary path is a factor of $4/\pi = 1.3$ times longer than the nominal point-to-point distance, and the flux is about 0.78 times the ideal capacity.

Combining these three flux decreasing factors, we can predict a corrected rule-of-mixing diffusivity for realistic polycrystals. Specifically, $D_{eff} \approx 0.27 D_b\delta / \bar{r}$, which is plotted as the lower line in Figure 3. The agreement between the corrected rule-of-mixing diffusivity and the observed effective diffusivity is excellent. In fact, the rule-of-mixing diffusivity is very slightly lower than the observed diffusivity, because it ignores bulk diffusion around the flux bottlenecks, which does occur in the simulations, particularly at small grain sizes. In addition, deviations from this scaling occur when the grain size approaches the same order of magnitude as our finite lattice.

TRANSIENT DIFFUSION PROFILES

The ability to closely observe transient diffusion is an advantage of the finite difference method. For example, Figure 4 shows solute concentration across a large-grain two-dimensional polycrystal during transient diffusion. The grain boundaries are clearly dilineated by solute concentration peaks, and concentration is relatively depleted near the grain centers.

Levine and MacCallum have proposed a model for solute penetration into a polycrystal [7]. They coupled transverse diffusion into the sample via a given volume fraction of high diffusivity boundary material with lateral bulk diffusion into an average-shape grain. Numerical analysis of their results predicts that the logarithm of the average concentration at depth x should scale with $x^{6/5}$ in certain time and penetration regimes. This result has become the basis for experimental determination of grain boundary diffusion coefficients. Solute concentration is measured in serial sections of a polycrystalline diffusion specimen and compared to the penetration curves of Levine and MacCallum [7].

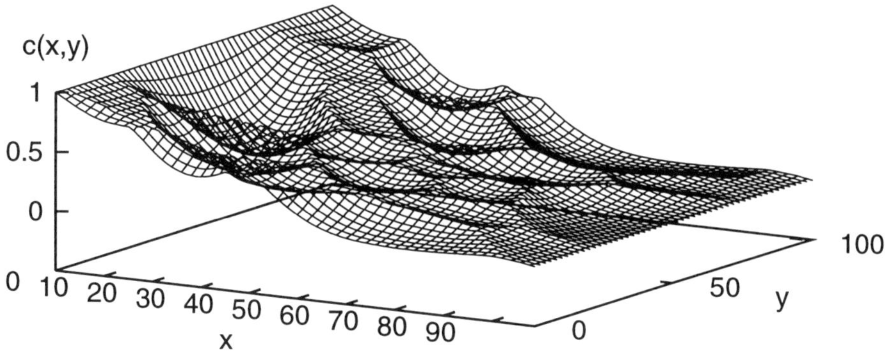

Figure 4. Solute concentration profile in a two-dimensional microstructure.

Solute penetration curves for finite difference diffusion through two-dimensional grain structures are shown in Figure 5. Curves show a fairly linear dependence of $log<c>$ on $x^{6/5}$ for small x, as predicted by Levine and MacCallum. The curvature at long times and large penetration depths is of the correct sign as well. The slopes of the linear regions of the curves are uniformly lower than the predicted slopes; that is, the solute concentration at a given penetration is smaller than predicted. There are two reasons for this solute deficiency. First, Levine and MacCallum combined all grain boundaries into a single volume fraction of boundary material, so bottlenecks at three-grain junctions and the path length effects discussed above are ignored. As in the steady state example, these effects will decrease the solute carrying capacity of boundaries in real polycrystals. Second, the stereological parameters used in the analysis of Levine and MacCallum are three-dimensional. The polycrystal in our study is strictly two-dimensional. Again, this decreases the solute carrying capacity of our boundaries. Nonetheless, solute penetration in transient state diffusion agrees well with the functional form given by analytical results.

CONCLUSIONS

We have implemented a flexible finite difference technique for solute diffusion in a bitmapped microstructure based on the Crank-Nicholson scheme that uses the Alternating Direction Implicit (ADI) method to provide stability for long timesteps. Our implementation uses a variable timestep to maintain a desired precision for both short-time and long-time regimes of diffusion.

We find excellent quantitative agreement with analytic results for one-dimensional diffusion problems, including the thin film solute source and semi-infinite slabs of differing diffusivities.

In simulations of steady-state grain boundary diffusion in two-dimensional polycrystals, we find that the effective diffusivity varies with the inverse of average grain size, as predicted by a rule-of-mixing diffusivity. Flux bottlenecks at three-grain junctions and geometric factors which increase boundary path length decrease the magnitude of the observed diffusivity to about 0.27 times the rule-of-mixtures prediction.

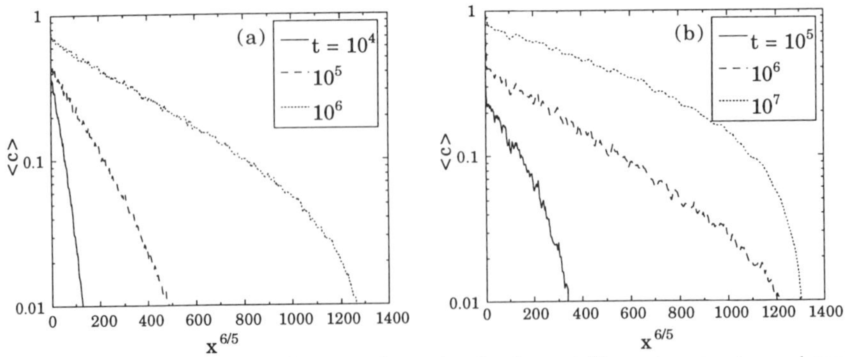

Figure 5. Solute penetration into two-dimensional polycrystalline microstructures plotted in the functional form suggested by Levine and MacCallum [7]. $D_g = 10^{-6}$, $D_b = 10^{-1}$, (a) $\bar{r} = 5$, (b) $\bar{r} = 10$.

Transient state boundary diffusion agrees with the solute penetration analysis of Levine and MacCallum [7]. However, the magnitude of the concentration as a function of penetration is again decreased due to junction bottlenecks and geometric factors.

Greater resolution is required to model thinner grain boundaries. Real microstructures contain grain boundaries with thicknesses that are at most 10^{-3} of the average grain diameter. In our simulations, grain boundary thicknesses are at least 10^{-3} of the average grain diameter. Thus, more realistic simulations will require the use of larger diffusion lattices.

ACKNOWLEDGMENTS

The authors appreciate the help of Michael F. Young and Stephen A. Wright in developing the diffusion codes. This work performed at Sandia National Laboratories, supported by the U.S. Department of Energy under contract number DE-AC04-94AL85000.

REFERENCES

1. I. Kaur, W. Gust, *Fundamentals of Grain and Interphase Boundary Diffusion*, 2nd ed. (Ziegler Press, Stuttgart, 1989).
2. J.C. Fisher, *J. Appl. Phys.* **22**, 74 (1951).
3. R.T.P. Whipple, *Phil. Mag.* **45**, 1225 (1954).
4. G.H. Gilmer and H.H. Farrell, *J. Appl. Phys.***47**, 3792 (1976).
5. G.H. Gilmer and H.H. Farrell, *J. Appl. Phys.***47**, 4373 (1976).
6. Y.M. Mishin, *Phys. Stat. Sol. A* **133**, 259 (1992).
7. H.S. Levine and C.J. MacCallum, *J. Appl. Phys.* **31**, 595 (1960).
8. J.P. Levine and D.L. Lossee, *J. Appl. Phys.* **56**, 924 (1983).
9. P.H. Holloway, D.E. Amos, and G.C. Nelson, *J. Appl. Phys.* **47**, 3769 (1976).
10. X. Gui, S.K. Dew, M.J. Brett, and D. de Cogan, *J. Appl. Phys.* **74**, 7173 (1993).
11. J. Crank, *The Mathematics of Diffusion*, 2nd. ed. (Clarendon Press, Oxford, 1975).
12. M.P. Anderson, G.S. Grest, and D.J. Srolovitz, *Phil. Mag. B* **59**, 293 (1989).
13. E. A. Holm, J. A. Glazier, D. J. Srolovitz, and G. S. Grest, *Phys. Rev. A* **43** 2662 (1991).

COMPUTER-ASSISTED MODELING OF HILLOCK GROWTH

Nicholas D. Rosen, The Pennsylvania State University,
129 Intercollege Materials Research Laboratory, University Park PA 16802

Susmit Kumar, The University of Cincinnati,
Department of Materials Science and Engineering,
497 Rhodes Hall, P.O. Box 210012,
University of Cincinnati, Cincinnati OH 45221-0012

Raj N. Singh, The University of Cincinnati,
Department of Materials Science and Engineering,
497 Rhodes Hall, P.O. Box 210012,
University of Cincinnati, Cincinnati OH 45221-0012

Stewart K. Kurtz, The Pennsylvania State University,
187 Materials Research Institute, University Park PA 16802

ABSTRACT

Hillock growth in aluminum interconnects is a major concern of the electronics industry. Using a grain map and orientation data from an actual Al film,[1] the authors have applied a finite element model to calculate the microstresses along individual grain boundaries, and combined these stresses with data on the variation of the grain boundary diffusion coefficient with misorientation angle to estimate relative diffusion rates and predict sites of hillock formation. The predictions of the model are then compared with the observed sites of actual hillock formation.

INTRODUCTION

Because aluminum has a higher coefficient of thermal expansion than silicon, aluminum interconnects in microelectronic devices are subject to hillock formation when heated, as occurs during fabrication and, in combination with electromigration, during use. Hillocks do not form at random locations, but at triple points where grains meet, [2,3] strongly suggesting that grain boundary diffusion plays a major role in their growth. It has also been observed that hillocks form on grains which do not share the usual <111> fiber orientation of aluminum films, and it is believed that the high resolved glide stresses in the <111> planes of misoriented grains produce local stress relaxation by dislocation glide, and the subsequent migration of atoms from high stress regions to these relaxed sites, where they accumulate in hillocks.[4,1]

Different models of hillock growth have been proposed. Although diffusion along the grain boundaries is widely accepted as the principal mass transport mechanism, Chaudhari argues that film-substrate interface diffusion best explains the results of his experiments with metal films on quartz surfaces.[5] Chang and Vook make the case for diffusion over the film's growth surface.[6] The three proposed mechanisms are illustrated in Figure 1.

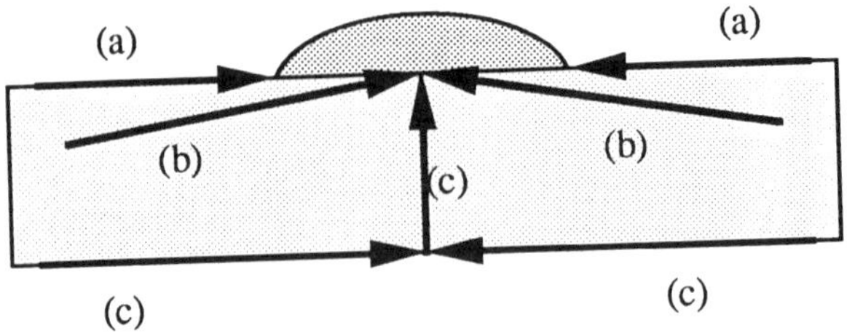

Figure 1. Three possible mechanisms of hillock growth are shown. (a) Surface diffusion, (b) grain boundary diffusion, and (c) film-substrate interface diffusion .

THE FINITE ELEMENT MODEL

The present authors have attempted to apply a two dimensional finite element model to the calculation of microstresses within an actual film. It has been established that the calculated stresses at triple points tend to either diverge or go to zero when a finite element model is used to simulate stress in a regular array of hexagonal grains,[7,8,9] and the same has been found for the more general case of irregular grains.[10] The question to be determined was whether hillock sites could be predicted from the distribution of microstresses in the simulated film. The simulation was based on the map of the grains in an actual annealed aluminum film, labeled film D1A1G, which Schwarzer and Gerth characterized by transmission electron microscopy.[1] Detailed information on the grain orientations was provided by Gerth.[11]

The finite element model divided each grain into a number of quadrilateral and triangular elements, and applied a 1.1% biaxial strain to the film. Each two dimensional grain in the model was assigned an orientation derived from that of the actual three dimensional grain as follows: A rotation was performed to bring the nearest <111> axis of the actual grain into exact correspondence with the normal to the plane of the film. (As the film had a <111> fiber orientation, the required rotation was normally small.) The same rotation was applied to the perpendicular <110> axes, and resulting orientation of the <110> axis nearest the x-axis of the film coordinate system was chosen as the two dimensional orientation of the grain.

The finite element package NASTRAN then calculated the resulting stresses in each element. The principal stresses σ_{xx} and σ_{yy} were always nearly equal to -1.84 GPa; the shear stress τ_{xy} was much lower, typically less than 10^{-3} GPa, and varied, being either positive or negative. It was noted that the simulation to some extent mirrored the tendency for stresses to diverge at triple points in the absence of plastic flow; the standard deviation of the shear stress was 8.48×10^{-3} GPa for all elements, but 2.44×10^{-2} GPa for elements at triple points.

For each element along a grain boundary, the stresses were transformed according to the rotation equation

$$\sigma_{ij} = a_{ik} a_{jl} \sigma_{kl}, \qquad (1)$$

to compute the shear stress along the direction of the grain boundary. Then the shear stress in each element was averaged with that of the corresponding element in the neighboring grain (see Figure 2). Thus, each of the n pairs of elements along a grain boundary had its associated shear stress, τ, and the mass flow to a triple point was taken as proportional to the difference between the average shear stress along the boundary and that in the elements closest to the triple point. The rate of mass flow J to the triple point next to the first pair of elements was taken to be

$$J \propto \frac{1}{n}\left(\sum_{i=1}^{n} \tau_i\right) - \tau_1, \tag{2}$$

and the mass flow to the triple point at the other end of the grain boundary was calculated similarly, but with τ_n in place of τ_1. The total mass flow to a triple point was the sum of the flows along the boundaries meeting there.

The mass flows to the triple points surrounding actual hillock sites were then compared to the mass flows to all triple points. Results are given in Table I.

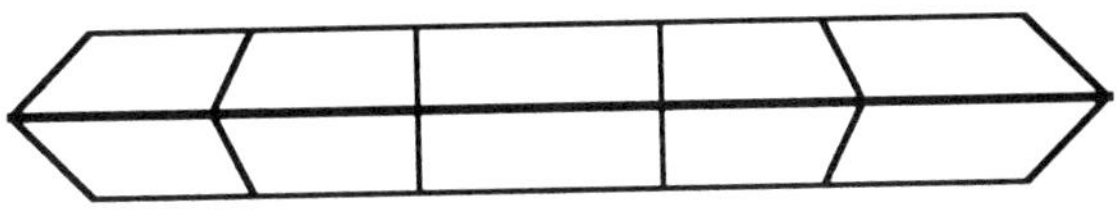

Figure 2. The elements on either side of a grain boundary were quadrilateral, with nodes in common as shown. Thus it was possible to group them in pairs, and average the shear stresses within each pair.

Table I. Calculated Average Mass Flows

	Arbitrary units:
To all triple points:	0.900
To triple points surrounding hillock sites:	0.773

It may be seen that the model did not successfully predict hillock sites; mass flows to the triple points around hillocks were actually calculated as being lower, on average, than mass flows in general.

STRESS RELAXATION

In a variation of the calculation described in the previous section, calculated shear stresses were modified to take stress relaxation into account. The grains with the highest resolved shear stresses along slip directions, which were also the grains beneath hillocks, were taken to have relaxed, and it was assumed that no mass flow took place along the boundaries of those grains. Mass flow on other grain boundaries terminating in triple points on relaxed grains was modified by setting the shear stresses to zero in the elements nearest those triple points, if the shear stresses in question were positive. They were left unchanged if negative. Equations (2) and (3) were applied as before, and the results are given in Table II.

Table II. Calculated Average Mass Flows with Stress Relaxation

	Arbitrary units:
To all triple points:	0.909
To triple points surrounding hillock sites:	1.047

Unsurprisingly, this modification of the simulation has a visible effect on the results, causing the model to predict more mass flow to the triple points around hillock sites than to other triple points. The effect is more dramatic if negative as well as positive shear stresses are modified. But whether that is done or not, the computer simulation of stress relaxation adds nothing basically new to the work of Schwarzer and Gerth.

GRAIN BOUNDARY DIFFUSION COEFFICIENTS

Several researchers have found that grain boundary diffusion coefficients in aluminum and other materials depend on the type and angle of the misorientation. This is the case for aluminum self-diffusion,[12] and for the diffusion of zinc in aluminum.[13,14] Only Biscondi[14] and his colleagues appear to have published graphs of diffusion coefficient as a function of twist and tilt angle, and so their results have been employed. The relative values of the grain boundary diffusion coefficients were calculated on the following assumptions. First, that diffusion coefficients for aluminum self diffusion are in proportion to those for zinc diffusion in Al; second, that the ratios between diffusion coefficients for different misorientations do not change with temperature; third, that grain boundaries are perpendicular to the plane of the film; fourth, that tilt boundaries were <110>; and fifth, that twist boundaries were <100>. All of these assumptions are to some degree inaccurate or questionable, but in the absence of complete data, it was necessary to make approximations.

Estimated diffusion coefficients for grain boundaries leading toward hillock sites were then compared to the coefficients for all grain boundaries, and to those for grain boundaries surrounding hillock sites. In each case, coefficients for diffusion within the plane of the film were of interest. Results are given in Table III.

Table III. Estimated Grain Boundary Diffusion Coefficients

	10^{-7} cm^2/sec:
For all grain boundaries:	2.322
For grain boundaries leading to hillock sites:	2.772
For grain boundaries bordering hillock sites:	1.784

According to a T-test, the increase in diffusion coefficients for boundaries leading to hillock sites was not statistically significant at the 5% level, although the decrease for boundaries bordering such sites was. It is therefore not established, although it remains possible, that the high grain boundary diffusion coefficients of certain misorientations played a role in determining hillock sites.

CONCLUSIONS AND FUTURE WORK

The finite element simulation could not in itself predict the sites of hillock formation. This may reflect the limitations of computer capacity that precluded a finer mesh, and surely reflects the fact that the stress relaxation of misoriented grains must be taken into account.

Currently available data relating grain boundary diffusion coefficients to grain misorientations are inadequate to allow mass flows to be predicted accurately.

To reduce hillock formation and improve the reliability of microelectronic devices, it is desirable to avoid depositing grains whose orientations will subject them to high resolved shear stresses along glide directions in slip planes. It may also be important to avoid misorientations which give rise to high grain boundary diffusion coefficients.

Several avenues for future research suggest themselves. With more computer capacity, it should be possible to employ a finer finite element mesh, to use elements with nodes along the edges as well as at the vertices, and to make the simulation three dimensional. It has been found that the accuracy of a finite element simulation of the stress distribution in one grain depends on the number of elements used;[10] it remains to be seen how well a more thorough simulation of the microstress distribution in a film will enable hillock sites to be predicted.

One of the authors, in collaboration with other researchers, is currently pursuing an experimental approach to the problem of distinguishing among mass transport mechanisms in thermal strain-induced hillock formation. Aluminum films have been deposited in layers, the different layers containing different impurities, and then annealed to produce hillocks. It is hoped that analysis of the hillocks and other sites by Auger spectroscopy and secondary ion mass spectroscopy (SIMS) will determine the origin of the material in the hillocks: the film's growth surface, its interior, or the film-substrate interface. Another project of interest is to use scanning electron microscopy to obtain the orientations of all aluminum grains in a region of film, then to predict hillock sites on the from the resolved shear stresses to which it is calculated that the various grains are subject, and from the estimated diffusion coefficients of the grain boundaries. The film would then be annealed, and actual hillock sites compared with those predicted.

ACKNOWLEDGMENTS

The principal author gratefully acknowledges the financial support of the Exploratory and Foundational Program of the Applied Research Laboratory at the Pennsylvania State University. The work of two of the other authors (Kumar and Singh) was made possible by National Science Foundation grant DMR-9202111.

1. R. A. Schwarzer and D. Gerth, "The Effect of Grain Orientation on the Relaxation of Thermomechanical Stress and Hillock Growth in Al-1%Si Conductor Layers on Silicon Substrates," *J. Elect. Mater.*, **22**(6), pp. 607-10 (1993).
2. E. Philofsky, K. Ravi, E, Hall, and J. Black, "Surface Reconstruction of Aluminum Metallization -- A New Potential Wearout Mechanism," pp. 120-28 in *Proceedings of the IEEE International Reliability Physics Symposium*, p. 120, IEEE, New York (1971).
3. D. Gerth, D. Katzer, and M. Krohn, "Study of the Thermal Behaviour of Thin Aluminium Alloy Films," *Thin Solid Films*, **208**, pp. 67-75 (1992).
4. J. E. Sanchez and E. Arzt, "Effects of Grain Orientation on Hillock Formation and Grain Growth in Aluminum Films on Silicon Substrates," *Scripta Metall. et Mater.*, **27**(3), pp. 285-90 (1992).
5. P. Chaudhari, "Hillock Growth in Thin Films," *J. Appl. Phys.*, **45**(10), pp. 4339-46 (1974).
6. C. Y. Chang and R. W. Vook, "Thermally Induced Hillock Formation in Al-Cu Films," *J. Mater. Res.*, **4**(5), p. 1172-81 (1989).
7. H. A. McKinstry, H. E. Schull, and W. R. Buessem, "Localized Stress in Elastic Polycrystalline Materials," pp. 695-705 in *Computer Simulation for Material Applications*, edited by R. J. Arsenault, R. J. Beeker, R. J. Beeker, and J. A. Simmons, Nuclear Metallurgy, Vol. 20, part 2, National Bureau of Standards, Gaitherburg, Maryland (1976).

8. --------, "Stress and Strain Fields in Idealized Polycrystalline Materials and the Prediction of Fracture Initiation Sites," pp. 929-41 of the same.
9. V. Tvergaard and J. W. Hutchinson, "Microcracking in Ceramics Induced by Thermal Expansion or Elastic Anisotropy," *J. Amer. Cer. Soc.*, **71**(3), pp. 157-66 (1988).
10. S. Kumar and S. K. Kurtz, "Micro-Stress Distribution within a Grain: Effect of Elastic Anisotropy," unpublished work.
11. D. Gerth, personal communication (1994).
12. J. Kwiecinski and J. W. Wyrzykowski, "Investigation Grain Boundary Self-Diffusion at Low Temperatures in Polycrystalline Aluminium by Means of the Dislocation Spreading Method," *Acta Metall. et Mater.*, **39**(8), pp. 1953-58 (1991).
13. W. Lojkowski, D.Beke, J. Kwiecinski, J. Wyrzykowski, and M. W. Grabski, "Application of the Compensation Relationship to Grain Boundary Diffusion in Aluminium and Interaction of Dislocations with Grain Boundaries," *Defect and Diffusion Forum*, Vol. **95-98**, pp. 475-478 (1993).
14. M. Biscondi, "Intergranular Diffusion and Grain-Boundary Structure," pp. 225-241 in *Physical Chemistry of the Solid State: Applications to Metals and Their Compounds*, edited by P. Lacombe, Elsevier, New York (1984).

Fracture

FRACTAL ANALYSIS OF GLASS AND CRYSTALS USING MOLECULAR DYNAMICS

Y.L. Tsai, T. Swiler, J.H. Simmons, and J.J. Mecholsky, Jr.
Department of Materials Science and Engineering
University of Florida, Gainesville, FL 32611

ABSTRACT

We present a simulation of fracture in silica glass and single crystal silicon using molecular dynamics. A fracture surface is then created by assuming a spherical electronic cloud. Sections of the surfaces create slit island contours from which fractal dimensions can be obtained. The fractal dimensions obtained from the simulated fracture surfaces agree well with those measured from fractured flexure bars.

INTRODUCTION

The understanding of the fracture process in brittle materials entails an understanding of bond breaking processes as well as macroscopic manifestations such as characteristic features on the fracture surface.[1] On brittle fracture surfaces, these features have become known as mirror, mist and hackle and are a manifestation of the crack path deviating from the plane.[2] It is difficult to be able to make measurements at all length scales. Modeling of the fracture process aids in the understanding of the process if the appropriate assumptions are made. The long range goal of our study is to understand the fracture process from the breaking of bonds to the process of macroscopic crack branching. The ideal situation would be to be able to model the process at all length scales and to compare modeling with the experimental results. Since we are not able to perform this analysis at this time, the next best approach is to select appropriate models at whichever length scales are possible and to compare to experimental results.

Our present approach is to model the fracture of an inorganic (silica) glass and a (silicon) single crystal at the atomic scale using molecular dynamics simulations and compare the results to experimental measurements of the surface topography. Since we do not have reliable direct measurement of the bond breaking process at the atomic scale, we take advantage of fractal geometric scaling algorithms to compare results at the atomic length scale to those at the macroscopic length scale. The tenets of fractal geometry assert that a process such as bond breaking will repeat and magnify as it increases in extent and with time. The assumption is that fracture in these brittle materials is a fractal process both in time and space. If this is true, then it should be possible to find the scaling rules that govern the process. The analyses described here is our first attempt at understanding the fracture process using this approach directly. What we describe here is a simulation of fracture at the crack tip using molecular dynamics (MD). We then create a fracture surface assuming that the electron cloud will form a spherical surface around each of the atoms. The fractal dimension of the formed surface will then be determined using conventional techniques, i.e., slit island analysis. The fractal dimension will then be compared to that measured on fracture surfaces of the same materials using slit island analysis from replicas of the macroscopic surface and box dimensions from atomic force microscopy measurements. The underlying assumption is that the process occurring at one portion of the crack tip is repeated and magnified during crack propagation. The dynamic effects, which must occur, are accounted for during the process and are related to the initiation of fracture.

First we will describe the MD simulations in silica glass and silicon single crystal. Then we will describe the method of fractal analysis for the MD simulations and for the experimentally obtained fracture surfaces. Finally, we will present the results, discuss the implications of the comparisons of the measurements, and conclude with a summary.

ANALYTICAL AND EXPERIMENTAL PROCEDURES

In this paper, the fracture dynamics and surface reconstruction of glass and of single crystal silicon are examined using MD simulations. Several aspects of the MD approach used in this paper must first be discussed, including the force functions, boundary conditions, and methods for stress or strain application. The MD computational algorithm used is based on the Verlet algorithm which has been widely used in studies of glass and crystal structure and dynamics, including the fracture of silicon dioxide.[3,4]

The force function used to model the ionic interatomic bonding of SiO_2 is the Born-Mayer-Huggins potential, shown below.[3,5]

$$\Phi_{ij} = A_{ij} \cdot \exp\left(\frac{-r}{\rho}\right) + \left(\frac{q_i q_i}{r}\right) \cdot \mathrm{erfc}\left(\frac{r}{\varepsilon L}\right) \tag{1}$$

where $\mathrm{erfc}\left(\frac{r}{\varepsilon L}\right)$ is the complement error function, ε is a constant, L is a linear dimension of the molecular dynamics cell, q is the charge, r is the distance between the i and j atoms, and ρ is a hardness parameter. The details of the modeling are presented elsewhere.[3,4]

The force function used to model the covalent interatomic bonding of Si is that of Stillinger and Weber (SW).[7] The function includes angular-dependent terms which model the directionality of covalent bonding. This force function has had great success in producing tetrahedral bonding in the structure of Si, as well as fitting its elastic and thermodynamic (melting) properties.[7,8] There are many references to demonstrate the usefulness of this potential.

Although the Stillinger-Weber (SW) potential has been shown to successfully simulate bulk Si properties, many problems arise at a free surface. At fracture, the SW potential exhibits an attractive force with too large a range leading to the formation of unphysical threads consisting of single Si atoms aligned together, each having a coordination number of 2, connecting the fracture surfaces. Since the modeling of fracture depends strongly on the reconstruction of the new surfaces, a more reasonable surface simulation is required to obtain any realistic results.

Examinations of the fracture process with the SW potential at the surface showed that the long-range attractive component of the potential allowed bonding to persist long past separation of the two faces. This resulted in the formation of the long threads or fibers discussed above connecting the fracture faces. This behavior clearly shows that the use of the unmodified SW potential at the fracture surface is inappropriate.

While a change to a more appropriate potential is necessary at this point, it was clear that abruptly introducing a new function could jeopardize any interpretive value of the studies. Instead, a physically supported modification was implemented as described below.

It is well known that a significant amount of particle, charge, phonon and photon emission occurs during fracture at various materials, including Si.[13] Therefore, without considering the exact processes of breaking a covalent bond, the analysis presented assumes that the bonding charge is emitted during bond fracture. Modeling of this process can be implemented by maintaining a SW potential interaction between Si particles up to a maximum distance corresponding to the cut-off distance in the potential function. Beyond this point, it is assumed that the bonding electrons are emitted, imparting a positive charge to the remaining ions. Thus, Si ions with four neighbors are uncharged, while those with fewer neighbors acquire a charge corresponding to the missing neighbors. This charge redistribution suggests the implementation of a repulsive Coulomb-like interaction between under-coordinated Si atoms after fracture. While this is only one of the several modifications which could have been implemented to produce a more reasonable fracture surface, we propose that it has the best physical basis.

This approach is supported by the results of fracto-emission studies[13] and by the lack of thread formation after fracture in Si. It is also a reasonable result of distorting a highly directional bond. The assumption is consistent with the high activity of fractured Si surfaces and their great affinity for oxygen and hydroxyls.

This approach is expected to slightly over-estimate the effect of bond breakage in instances where covalent bonds break and reform before the electrons are ejected. The approach also is likely to over-estimate the concentration of highly charged Si ions with only a single neighbor, or totally dissociated ions. However, these structures occur so seldom in the simulation that the issue is inconsequential.

The assumption presented is critical to the study in that it allows the modeling of a covalent structure with an ionic free surface. The approach is novel and removes the problem of non-physical low coordination bonding inherent in the SW potential. Finally, the approach does yield a reasonable surface behavior for Si atoms and, as shown below, does lead to realistic surface geometries.

The formation of a Si-Si interionic force is best done along the lines developed for the silicon dioxide bonding studies of Soules[14] and Ochoa et al.[3,15] Here a reasonable simulation of the structure of silica has resulted from the selected Born-Mayer-Huggins (BMH) force functions as shown earlier. These studies have shown that the BMH potential simulates well the ionic interaction with a minimum of adjustable parameters. Furthermore, the choice of the BMH potential allows the use of parameters already developed for the Si-Si ionic interaction in amorphous SiO_2, so that no new or untried parameters are introduced in this study.

Periodic boundary conditions were used across the 3 spatial axes. An energy distribution function is used to assign the energy (velocity) of each atom at the beginning. The temperature of the sample can be varied by scaling the velocity of each particle with an appropriate factor and letting the total kinetic energy of the particles be equal to the enthalpy of the desired temperature. The temperature of the sample was kept constant by scaling the velocity every time step to avoid heating (or cooling) effects. The details of the process are then similar to that described previously for silica.[3,4]

Fractal geometry is a non-Euclidean geometry which uses a non-integer dimension (the fractal dimension) and lacunarity to evaluate surface tortuosity and tests for self-similarity or self-affinity and scale invariance. The fractal dimension characterizes the condition of a surface and the self-repeating patterns necessary to replicate the surface topography. Previous work[16-20] has shown that there is a correlation between the fractal dimension and the fracture toughness in brittle materials. A fractal structure can be constructed by replication of a generator shape which is scaled by superposition on an initiator geometry. Presumably, if the fracture process and resulting fracture surfaces are fractal, as several experimental observations have implied, then it should be possible to determine the initiator and generator shape through modeling such as MD.

Fractal geometry is characterized by the fractal dimension and is related to measurement theory. The fractal dimension can be obtained from the simulated fracture surface using the slit-island technique.[21] The measured length of a line can be described by the following equation, called Richardson's equation:

$$L = kS^{1-D} \tag{2}$$

where L is the measured length, S is the measured scale, D is the dimension of the curve, k is a proportionality constant.

The fractal dimension of a line can be computed from the above equation. The line's length is computed over a range of scales. A log-log plot of length vs. scale gives a straight line with slope equal to 1-D.

Fracture surfaces were constructed from the simulations by assuming spherical electronic clouds with a radius equal to the cutoff distance of the potential. Contour plots can be obtained by cutting the simulated fracture surfaces at suitable heights. The contour at each height is similar to the experimental slit-island section. Fracture surfaces from four point flexure bars were obtained in a similar manner and the technique is explained in detail elsewhere.[21,22]

Atomic force microscope images can be analyzed in a similar manner as the contours of the experimentally obtained surfaces. However, because of the algorithm used, the box dimension is obtained instead of the contour dimension.

RESULTS AND DISCUSSION

Figure 1 shows the fracture surface obtained using MD simulations for silicon in the (100) plane. This is a representative reconstruction of all of the surfaces obtained. The details of the shape of the surfaces change depending on the orientation of the crystals, but the character is the same. Contour sections of the surfaces were obtained from the computer simulations and a representative section is shown in Figure 2. The fractal dimensions were then determined from

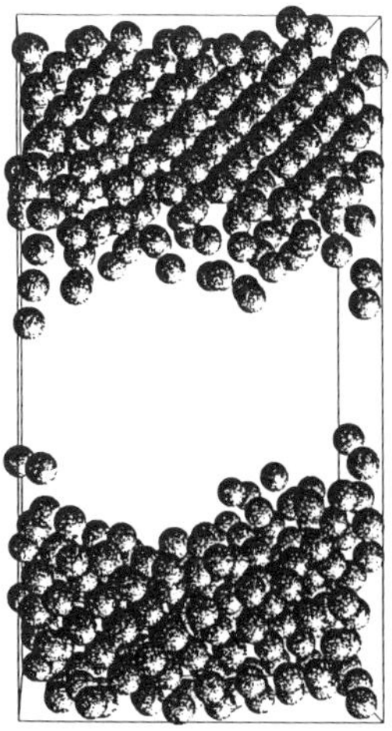

Figure 1. Atomic position simulation from MD fracture modeling. [Silicon [(100)]

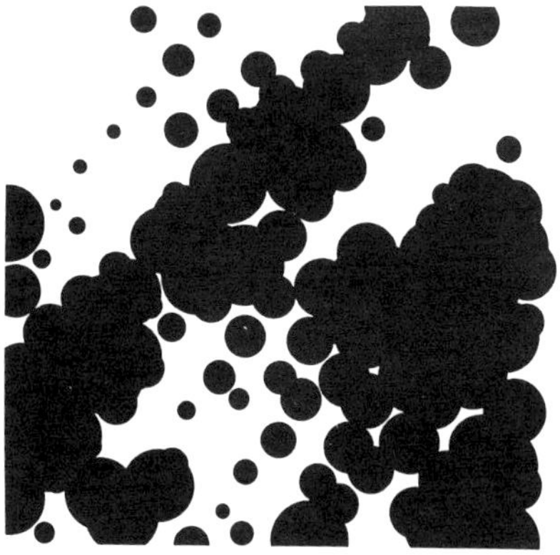

Figure 2. Slit island contour from computer simulation of atom positions, e.g., as in Figure 1. [Silicon (100)]

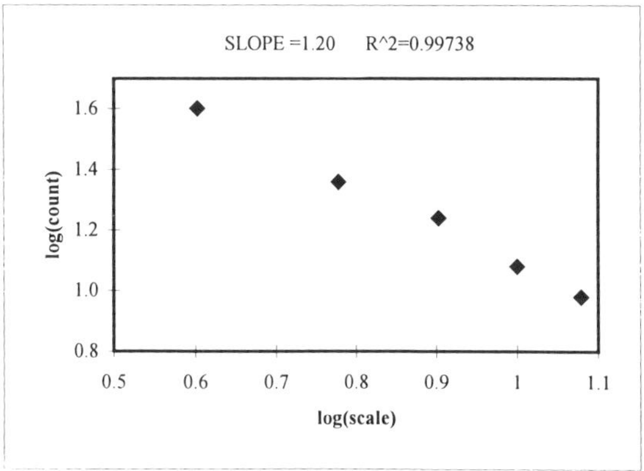

Figure 3. Graph of ln (number of counts) versus ln (scale length). This graph is the unitless representation of Eq. (2).

these contours by measuring the perimeter using different measuring scales. Thus, the log length versus log scale is graphed and the slope is -D* according to Eq. (2), where D* is the fractional part of the fractal dimension. An example graph is shown in Figure 3. The fractal dimensions were also obtained from measurements of the fracture surface on bars broken in four point flexure. Silica glass fracture surfaces were measured in this study and the silicon data were reported previously.[3,4] Table I compares the fractal dimensions calculated from the fracture simulations and the measured values from fractured bars of the material for silica glass and silicon tested in three orientations. The values are reasonably close. This implies that the fractal nature of the fractal process holds even to the atomic scale. This result is consistent with limited measurements of the fractal dimension on silica and silicon at other laboratories.[23,24]

Table I. Fractal Dimensions for Experimental and Simulated Fracture Surfaces

Material	Slit-Island	AFM	MD
Silica	2.10 ± 0.02	2.11 ± 0.01	2.11 ± 0.02
Silicon			
{100}	2.16 ± 0.04	2.19 ± 0.02	2.16 ± 0.06
{110}	2.10 ± 0.04	2.11 ± 0.02	2.11 ± 0.05
{111}	2.06 ± 0.02	2.05 ± 0.01	2.09 ± 0.04

We recognize that the MD simulations are limited in that the number of atom samples is small. In addition, the Stillinger-Weber potential is not necessarily the best one to select for representation of a silicon surface, nor is there any guidance in the literature for a better potential. However, the authors think that the combination of SW for the bulk and an ionic potential for undercoordinated atoms has provided a good starting point to attempt modification of the surfaces to reflect a condition closer to reality than the complete retention of covalent bonds, which is physically unrealistic. The BMH potential was selected because it satisfactorily described the fracture behavior of an ionic glass and crystal and has yielded a good Si-Si repulsion. Because of the comparison of the fracture surface simulation with experimental data of the same type, we have a way of analyzing the appropriateness of the selected potentials.

There is ample evidence that the brittle fracture surface of brittle materials are fractal.[20,21] The question remains to be answered to what length scale does this hold. Limited data[23,24] suggests that the fractal character is applicable even to the level of the atomic scale. If this is true, then a small portion of the surface should be representative of larger portions. Thus, it is reasonable for these cases to expect that a small representation of the crack tip can be extended to a greater portion of the crack tip. The selection of the potential for representation of the forces between atoms is important to the specific behavior of the atoms and molecules. By using this technique of comparison, we suggest that the different potentials can be tested for their appropriateness. This paper suggests a methodology by which this can be performed.

CONCLUSIONS

We simulated the fracture of silica glass and silicon single crystals using molecular dynamics techniques. We developed a technique whereby we can calculate the fractal dimension of the fracture surface from the simulations. These values compared well to experimental values. The implications of this work are that fracture surfaces are fractal at atomic length scales and can be modeled. Further, the results of the work imply that it may be possible to find an initiator and generator to produce artificially a fracture surface with characteristics which can be measured and compared to experimentally obtained fracture surfaces.

REFERENCES

1. J.J. Mecholsky, S.W. Freiman, and R.W. Rice, J. Mater. Sci., 11, 1310-1319 (1979).
2. J.J. Mecholsky, Jr., Ceramic Transactions 17, *Fractography of Glasses and Ceramics II*, 413-451 (1991).
3. R. Ochoa, T.P. Swiler, and J.H. Simmons, J. Non-Crystal. Sol., 128, 57 (1991).
4. J.H. Simmons, T.P. Swiler, R. Ochoa, J. Non-Crystal. Sol., 134, 179-182 (1991).
5. T.F. Soules, J. Non-Crystal. Sol., 123, 48-70 (1990).
6. F.H. Stillinger and T.A. Weber, Phys. Rev. B, 31[8], 5262 (1985).
7. S.R. Philpot, J.F. Lutsko, D. Wolf, and S. Yip, Phys. Rev. B, 40[5], 2831 (1989).
8. J.Q. Broughton and X.P. Li, Phys. Rev. B, 35[17], 9120 (1987).
9. G.J. Dienes and A. Paskin, J. Phys. Chem. Solids, 48, 1015-1033 (1987).
10. M. Mullins, Phil. Mag., 46[5], 771-787 (1982).
11. K.S. Cheung and S. Yip, Phys. Rev. Lett., 65[22], 2804-2807 (1990).
12. S. Kohlkoff, P. Gumbsch, and H.F. Fischmeister, Phil. Mag., 64[4], 851-878 (1991).
13. S.C. Langford, D.L. Doering, and J.T. Dickinson, Phys. Rev. Lett., 59[24], 2795-2797 (1987).
14. T.F. Soules, J. Non-Crystal. Solids, 73, 315 (1985).
15. R. Ochoa and J.H. Simmons, J. Non-Crystal. Solids, 75, 413 (1985).
16. Y.L. Tsai and J.J. Mecholsky, Jr., J. Mater. Res., 6[6], 1248 (1991).
17. J.J. Mecholsky, Jr., T.J. Mackin, and D.E. Passoja, Adv. in Ceram. 22, *Fractography of Glasses and Ceramics*, The American Ceramic Society, (1988).
18. T.J. Mackin and J.J. Mecholsky, ARL Technical Report No. TR87-007, Applied Research Lab., Pennsylvania State University (1987). (Also cf. T.J. Mackin, Master's thesis, *A Fractal Analysis of Brittle Fracture*, Pennsylvania State University.)
19. D.E. Passoja, Adv. in Ceramics 22, *Fractography of Glasses and Ceramics*, The American Ceramic Society, 101-126 (1988).
20. Y. Fahmy, J.C. Russ, and C.C. Koch, J. Mater. Res., 6[9], 1856-1861 (1991).
21. B.B. Mandelbrot, D.E. Passoja, and A.J. Paulley, Nature 308, [5961], 721-722 (1984).
22. J.J. Mecholsky, Jr., D.E. Passoja, and K.S. Feinberg-Ringel, J. Am. Ceram. Soc., 72[1], 60-65 (1989).
23. M.W. Mitchell and D.A. Bonnell, J. Mater. Res., 5[10], 2244 (1990).
24. D.M. Kulawansa, L.C. Jensen, S.C. Langford, J.T. Dickinson, and Y. Watanabe, J. Mater. Res. (1994).

Computational Models for Brittle Fracture

A. Jagota[1] and S.J. Bennison[2]

[1] Department of Applied Mechanics, Indian Institute of Technology, Delhi, Hauz Khas, New Delhi 110016, INDIA.
[2] Central Research and Development, E.I. DuPont de Nemours & Co., Inc., Wilmington, Delaware 19880-0356, USA.

Abstract

Issues relating to the development of quantitative computational models for brittle fracture of disordered microstructures are discussed. We examine the use of spring-network (SN) and finite-element (FE) models where fracture is simulated by the removal or fragmentation of elements or element boundaries. Some basic questions regarding the validity of such models for fracture that have only recently been raised are considered. SN models are not always able to capture a state of uniform stress. The use of local element breaking rules based on critical stress or energy leads to Griffith scaling of fracture strength, provided a length scale is included in the breaking rule. However, the perfect strength of the model is then far too low; this leads us to propose that different failure rules be used for elements at a crack tip and those removed from it. Regular meshes and networks are highly anisotropic with respect to crack propagation. Random meshes that obviate this problem introduce some randomness in the scaling of fracture strength with crack length.

1. Introduction

In many natural and synthesized materials, the fracture process consists of multiple, interacting cracking events. These are intimately related to heterogeneities in the microstructure of the material. Often, an analysis based on the mechanics of a single crack is inadequate. This has lead to the development of computational models for the fracture of heterogeneous materials which fall into two broad categories: (a) spring-network (SN), and (b) finite-element (FE).

By SN we mean the class of models in which a network of one-dimensional elements is used to represent a continuum of two or three dimensions. Various element types, *e.g.*, central-force springs, bond-bending springs, and beam elements have been used [1-9]. Failure is modeled by sequentially removing such elements as exceed a threshold value, typically of stored energy or force. By FE we mean the class of models in which the continuum governing equations are discretized over finite elements or volumes [10-13]. Failure can be modeled in number of ways: removal of elements [1,2], separation of element boundaries [12], or element fragmentation [13]. These methods differ from the conventional computational analyses of fracture in which the focus is on individual crack tips. They are attractive because of their simplicity and their ability to model structures with a relatively large number of microstructural features and cracks. SN methods have been used widely to study general scaling behavior in the fracture of brittle materials [4,6,8], and the fracture of different types of materials [see references in 1-3]. FE methods have been used most widely in the conventional sense, and recently in the present context [1,2,12,13].

However, several basic issues regarding the validity of SN and FE models for this sort of analysis have only recently been raised [1-3,7-9]. In this respect, we emphasize that for special problems such as the deformation and fracture of granular materials, where a direct correspondence between the discrete SN model and the microstructure can be established, SN models are valid *a priori*. Our concern here is with the broader class of problems where there is no relationship between the discrete length scales of the SN or FE model and the material. Then, the computational model is intended as a discretization of the continuum. We have earlier proposed [1,2] to examine these computational models by studying their performance as models for an isotropic, linearly-elastic, and brittle material in two dimensions. The specific questions we seek to answer are:

(a) *Are the models valid discretizations* ?

Because large spring networks have effective elastic properties, it is usually argued that the model is a discretization. We have argued in [1] that this homogenization argument should not be employed to justify a model as a discretization. It has also been shown that although a finite-volume discretization of the stored elastic energy resembles the discrete SN expressions, they are generally not identical. A numerical test that addresses this question has been proposed: the ability of the network to capture a uniform state of strain. Numerical results show that random spring-networks generally are unable to represent uniform strain, whereas arbitrary FE meshes are. Recently, it has been shown that with an appropriate assignment of spring properties, a beam SN can be constructed that can represent uniform strain [3].

(b) *What is the relationship between element breaking rules and macroscopic fracture toughness* ?

A related question is whether, with the given method of analysis, Griffith scaling of fracture strength $\sigma^{\infty c}$ with crack length c is obtained ($\sigma^{\infty c} \sim 1 / \sqrt{c}$ [14]). Especially for small cracks, it is clear that the models do not compute crack tip stresses with any accuracy. However, it can be shown that if a length scale characteristic of the mesh is included in the failure criterion, the scaling of fracture strength with crack length follows Griffith scaling for surprisingly small cracks. This issue is discussed in greater detail in section 2.

(c) *How does the model treat nucleation of cracks* ?

A mesh or lattice with no flaws has a perfect strength which is much smaller than the perfect strength of the material being modeled. This is unavoidable if the same failure criterion is used for elements at a crack tip and those removed from it.

The use of different failure criteria for elements depending on their location with respect to existing cracks raises new physical and algorithmic issues which are discussed in the section 3.

(d) *Is the model isotropic with respect to crack propagation* ?

It has been shown [1-3] that regular SN and FE meshes are highly anisotropic with respect to fracture; cracks propagate preferentially along lattice directions. Random meshes obviate this problem, but introduce some randomness in the strength-crack size scaling.

2. Crack propagation

As shown in [1], for any set of points, a unique spring-network can be constructed by connecting points that share an edge of the Voronoi cell diagram of the points. Also, a dual finite element mesh can be constructed using the same Voronoi cells. Part of such a construction is shown below in Figure 1; the thick lines representing the springs and the thin lines representing the Voronoi cells. Imagine this mesh, which has a crack consisting of a series of broken springs or elements, to be subjected to a remote stress σ^{∞}. In this section we discuss the relationship between local breaking rules and macroscopic fracture toughness. The discussion aims to establish scaling relationships between the two, independent of the specific type of SN or FE model employed.

Consider first the case of the spring-network. If a spring of length L* is removed when its stored energy exceeds e_c, it corresponds to a critical energy release rate:

$$G_c \sim e_c / L \qquad (1)$$

where L is the length of the Voronoi edge corresponding to the spring, and represents the amount the crack has advanced. Assuming that the *scaling* of local forces is that of a crack with a finite tip radius, one can estimate the force in the spring at the crack tip as:

$$F^{tip} \sim \sigma^{\infty} L \sqrt{(c / L^{*})} \qquad (2)$$

where c is the crack length. It follows that the energy in this spring, e^{tip}, scales as:

$$e^{tip} \sim F^{tip} \varepsilon L^{*} = (\sigma^{\infty})^{2} Lc / E \qquad (3)$$

Then, the fracture strength scales as:

$$\sigma^{\infty c} \sim \sqrt{G_c E / c} \qquad (4)$$

This shows that a failure criterion based on critical energy in a spring, normalized by the length of the Voronoi edge, yields Griffith scaling of failure strength with crack size. If L were not introduced into the failure criterion, the strength would still scale as the inverse square root of c. However, it would not be independent of the scale of the mesh; a larger mesh with the same length of crack would have lower strength. Because, in the scaling sense, breaking rules based on a critical value of force or strain correspond to a critical value of stored energy, the argument holds in those cases as well.

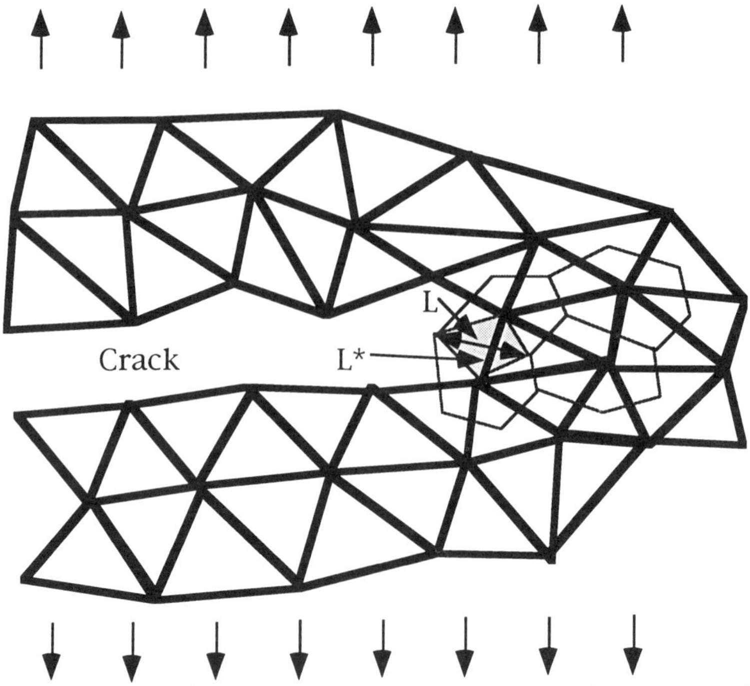

Figure 1: *Partial view of a crack in a spring-network or a finite-element mesh. The thick lines represent the spring-network. The thin lines represent the Voronoi cells of the nodal points which are used to construct the dual finite-element mesh.*

Several methods of crack propagation can be employed with FE methods, *e.g.*, crack removal, separation along element boundaries, and element fragmentation, as shown schematically in Figure 2. The case of crack removal has been discussed in [2], and as in the case of SN, it has been shown that a failure criterion based on stored energy, stress, or strain must include a length scale.

Consider the case of crack propagation by separation of element boundaries or by element fragmentation. Let us adopt a stress-based failure criterion:

$$\sigma^{tip}\sqrt{L} > K_c \qquad (5)$$

where K_C is the fracture toughness and σ^{tip} is an appropriate component of the stress. The scaling of this stress with crack length is:

$$\sigma^{tip} \sim \sigma^{\infty}\sqrt{(c / L)} \qquad (6)$$

Combining the two yields the scaling of fracture strength with crack size:

$$\sigma^{\infty c} \sim K_c / \sqrt{c} \qquad (7)$$

Again, it can be shown that this is equivalent to a criterion based on stored energy (divided by a length scale).

These scaling arguments show that most element breaking rules imply Griffith scaling of fracture strength with crack size. To make the scaling independent of the mesh size, a length scale characteristic of the element size must be included in the failure criterion. The major assumption in the analysis is that local forces or stresses scale with crack size as they would in a sharp crack. A numerical test of this assumption is shown in Figure 3.

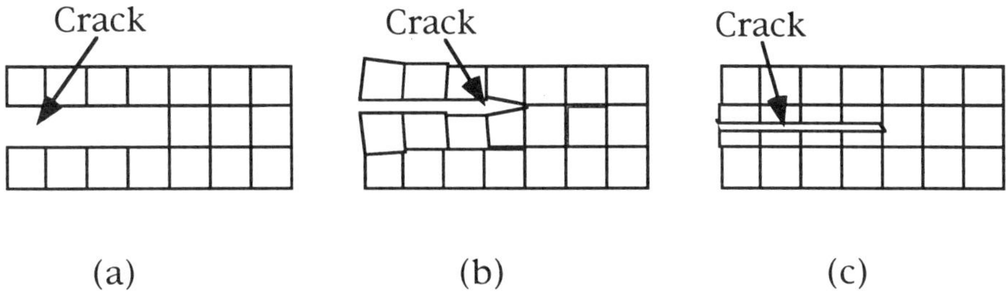

Figure 2 *Three methods of crack propagation in finite elements: (a) element removal, (b) separation of element boundaries, (c) element fragmentation.*

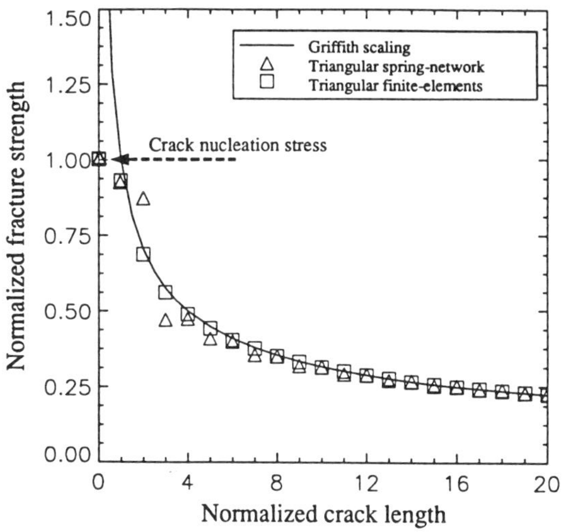

Figure 3 *Fracture strength normalized by the strength of the perfect mesh as a function of crack length.*

When the crack is more than 8-10 elements long, the strength scaling of the network/mesh matches Griffith scaling very well. For finite-elements, the scaling works for cracks as little as 3 elements long. Therefore, although the coarse discretization gives poor estimates of near-tip stresses, they nevertheless scale correctly with increase in crack size. This surprising result justifies the use of simple meshing and element breaking rules for fracture.

Note from Figure 3 that both the SN and FE models have a finite strength in the absence of a crack in the mesh. Clearly, an element breaking rule calibrated to yield the correct macroscopic fracture toughness also fixes a crack nucleation stress. The question as to whether the value of this nucleation stress is appropriate is addressed in the next section.

3. Crack nucleation

Consider a finite-element mesh without a crack subjected to a remote stress σ^∞, which is also now the uniform stress everywhere. If the crack propagation criterion is stress-based, then we immediately find that the nucleation stress is:

$$\sigma^n \sim K_c / \sqrt{L} \tag{8}$$

or equivalently,

$$\sigma^n \sim \sqrt{(GE / L)} \tag{9}$$

This is similar to the expression for the strength of a perfect crystal [14], if L were interpreted as the Burger's vector. An identical result is obtained for spring networks. Note also that if a length scale is not employed in the breaking rule, the ratio $(\sigma^n / \sigma^{\infty c})$: the ratio of nucleation stress to fracture strength with a crack of length c, remains unchanged.

Because L is typically much larger than atomic dimensions, σ^n seriously underestimates the nucleation stress. Moreover, the nucleation stress is mesh size dependent. This has several important consequences. Because σ^n is usually far too small, simulations of disordered materials are biased towards the nucleation of new cracks remote from the main crack. In disordered microstructures this can overestimate the formation of bridging elements. If irregular meshes are employed, artificial bridging can occur because a larger element away from a crack may fail before a small element at the tip, as shown in Figure 11 [2].

We therefore propose that different failure criteria be employed for elements at a crack tip and those removed from it. This, we think, is unavoidable if one is to match both macroscopic fracture toughness and a physically reasonable nucleation stress with an arbitrary mesh. It does introduce extra algorithmic complexity; one must maintain lists of crack-tip elements, and the list must be updated following each element break.

4. Conclusions

Element breaking rules in SN and FE models for brittle fracture yield Griffith scaling of failure strength with crack size. A length scale must be included in the element breaking rule if the failure strength is to be independent of mesh scale. The perfect strength of the material is seriously underestimated if the size of the elements is much greater than atomic dimensions. This can bias simulations towards predicting nucleation of cracks and bridge formation. Therefore, it is proposed that different failure criteria be used for elements at a crack tip (for crack propagation) and those removed from it (for crack nucleation).

5. References

[1] A. Jagota, S.J. Bennison, "Spring-network and finite-element models for elasticity and fracture", in *Non-Linearity and Breakdown in soft Condensed Matter,* Springer-Verlag Lecture Notes in Physics, v 437, 186-201 (1994), K.K. Bardhan, B.K. Chakrabarti, A. Hansen (editors).

[2] A. Jagota, S.J. Bennison, "Element breaking rules in computational models for brittle fracture", *Modelling and Simulation in Materials Science and Engineering*, in press (1995).

[3] E. Schlangen, E.J. Garboczi, "New method for simulating fracture using an elastically uniform random geometry lattice", submitted for publication to *International Journal of Engineering Science.*

[4] M. Sahimi, S. Arbabi, "Mechanical and electrical breakdown in disordered solids", *Physical Review B,* v 47 [2] 713-722 (1993).

[5] A. Hrennikoff, "Solutions of problems of elasticity by the framework method", *Journal of Applied Mechanics,* A169-175 (1941).

[6] D.J. Srolovitz and P.D. Beale, "Computer Simulation of failure in an elastic model with randomly distributed defects", *Journal of the American Ceramic Society,* v 71 [5] 362-369 (1988).

[7] W.A. Curtin, H. Scher, "Mechanics modelling using a spring network", *Journal of Materials Research,* v. 5 [3] 554-562 (1990).

[8] S. Roux, "Continuum and discrete description of elasticity and other rheological behavior", in *Statistical Models for the Fracture of Disordered Media*, H.J. Hermann and S. Roux (editors) 87-114, Elsevier Science publishers B.V (North-Holland) (1990).

[9] W.A. Curtin, H. Scher, "Brittle fracture in disordered materials: a spring network model", *Journal of Materials Research,* v 5 [3] 535-553 (1990).

[10] O.C. Zienkiewicz, R.L. Taylor, *The Finite Element Method,* 4th edition, McGraw-Hill (London) (1989).

[11] E. Onate, M. Cervera, O.C. Zienkiewicz, "A finite volume format for structural mechanics", *International Journal for Numerical Methods in Engineering,* v37 181-201 (1994).

[12] M. Ortiz, S. Suresh, "Statistical properties of residual stresses and intergranular fracture in ceramic materials", *Journal of Applied Mechanics,* v60, 77-84 (1993).

[13] D.L. Rice, E.C. Ting, "Fragmentation algorithm for finite element failure simulation and analysis", *International Journal for Numerical Methods in Engineering,* v 36, 3859-3881 (1993).

[14] B.R. Lawn, *Fracture of Brittle Solids,* 2nd edition, Cambridge University Press, New York (1993).

FINITE ELEMENT ANALYSIS OF RESIDUAL STRESSES IN FIBER-REINFORCED MATERIALS

Susmit Kumar and Raj N. Singh
Department of Materials Science and Engineering
University of Cincinnati
Cincinnati, OH 45221-0012

ABSTRACT

Residual thermal stresses in graphite/aluminum fiber-reinforced metal-matrix composites are studied using three-dimensional finite element modeling assuming the anisotropic thermal expansion coefficient of graphite fiber. The finite element results are compared with the two-cylinder model results of Vedula et al.[1] and the experimental results of Tsai et al.[2] In the square and hexagonal array arrangements of fibers, the possible sites for the crack initiation are the points where the fiber-matrix interface intersects the lines joining the fiber center to its first nearest neighbors.

INTRODUCTION

Metal-matrix composites have been widely used as structural materials due to their high toughness and strength in certain directions. Generally, in these composites the coefficients of thermal expansion of the fiber and matrix have a very large mismatch which results in a significant amount of residual stresses during cooling from the processing temperatures. The residual stresses may decrease or increase the elastic response of the composites and may affect their mechanical behavior.

In the present study, the three-dimensional hexagonal array and square array finite element models are used to calculate the residual thermal stresses in Pitch 55 graphite / A201 aluminum fiber-reinforced composite. The finite element results are compared with the two-cylinder model results of Vedula et al.[1] and the experimental results of Tsai et al.[2] Effect of fiber volume fraction on the distribution of hoop and radial stresses at the fiber-matrix interface is also investigated.

MODEL

Two types of fiber arrays - square and hexagonal - have been used to model the fiber-reinforced composites. In the longitudinal direction, the fiber axes have been assumed to be parallel and of equal lengths. In each type of array, the smallest possible parallelepiped periodic cell has been used in the present finite element study. In the square array model, only one fiber (more specifically only one-quarter of a fiber) has been used and all the edges in transverse cross-section are of equal length. In the hexagonal array model, two fibers have been used and the ratio AB : AD is 1 : √3.

A pre-processor has been written to divide the three-dimensional model of fiber-reinforced composites into finite 'elements.' The mesh has been refined near the fiber-matrix interface to accurately calculate the stresses at the interface (Fig. 1). Number of elements inside the fiber and in the matrix are varied depending on the fiber volume fraction. Mostly 20 node quadratic brick elements have been used in the analysis. Few 15-node quadratic triangular prisms have been used in the transition region.

For the boundary conditions, the nodes on the planes X = 0, Y = 0, and Z = 0 were constrained to have displacements in their respective planes only. The nodes on other three boundary planes were constrained to have displacements along the normals to their respective planes. Temperature of the model was raised by ΔT and ABAQUS finite element software version 5.3 was used to calculate the residual thermal stresses in the model.

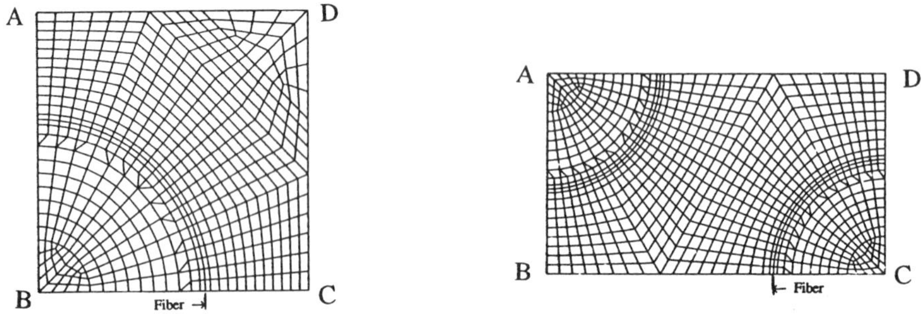

(a) Square Array (b) Hexagonal Array

Figure 1. Transverse 2D cross-sections of the three-dimensional models for the square and hexagonal arrays of fibers for 30 % fiber volume fraction.

RESULTS AND DISCUSSION

Table 1 shows the thermomechanical properties of aluminum and graphite. Fig. 2 shows the plot of maximum value of residual radial thermal stresses at the fiber-matrix interface for various fiber volume fractions for 1° C increase in temperature. For cooling, the radial stress will be - σ_r, i.e., compressive. In the square array model, the maximum radial stress increases with increase in the fiber volume fraction. In the hexagonal array model, it remains almost constant, and it decreases with the increase in the fiber volume fraction in the two cylinder model. Hence, the cubic array arrangement of fibers is more prone to cracking at the interface in comparison to the hexagonal array of fibers.

Table 1 Material Properties[1]

Material	Young's Modulus (GPa)	Poisson Ratio	Coeff. of Thermal Expansion - Transverse	Coeff. of Thermal Expansion - Longitudinal
Graphite (fiber)	385	0.20	25×10^{-6}	0
Aluminum (matrix)	70	0.35	23.6×10^{-6}	23.6×10^{-6}

Figure 3 shows the maximum value of tangential stress ($\sigma_{\theta m}$) in the matrix at the fiber-matrix interface, for various fiber volume fractions for 1° C increase in temperature. For cooling, the tangential stress is - $\sigma_{\theta m}$, i.e., tensile. For volume fractions < 0.4, the calculated maximum tangential stress for both the square array and hexagonal array models is less than those predicted by the two-cylinder model. For volume fractions > 0.5, the calculated maximum tangential stress for square array model is more than that predicted by the two-cylinder model.

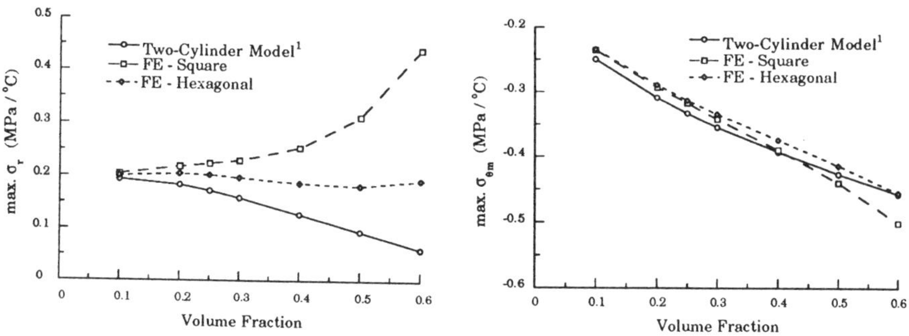

Figures 2-3. Maximum residual σ_r and σ_θ (in matrix) at the interface for 1° C increase in temperature vs fiber volume fraction.

Figures 4-5 show the residual radial and tangential thermal stress distributions at the interface for various fiber volume fractions for 1° C increase in temperature. For cooling, the sign of the stresses will change. Angles are measured anti-clockwise from the line BC and the line AD in square array and hexagonal array models, respectively (Fig. 1). The radial and tangential stress distributions have 45° and 30° symmetry in the square array and hexagonal array models, respectively. In both, the square array and hexagonal array models, the radial stresses on the fiber-matrix interface (Fig. 4) are maximum in magnitude at the points where the interface intersects the lines joining the fiber center to its first nearest neighbors (i.e. on the line BC for square array and 30°

anti-clockwise from the line AD for the hexagonal array in Fig. 1). In the square array arrangement of fibers, each fiber has four first nearest neighbors each at a distance of 2 x BC (Fig. 1), whereas in the hexagonal array arrangement of fibers, each fiber has six first nearest neighbors each at a distance of AC. These points on the interface are possible sites for crack initiation. Minimum radial stresses at the interface (Fig. 4) are at 45° anti-clockwise from the line BC in the square array and on the line AD for the hexagonal array model (Fig. 1). The maximum tangential or hoop stresses ($\sigma_{\theta m}$) on the interface (Fig. 5) are at the points of minimum radial stresses (i.e. at 45° anti-clockwise from the line BC in square array model and on the line AD in hexagonal array model). The minimum tangential or hoop stresses on the interface (Fig. 5) are at the points of maximum radial stresses (i.e. on the line BC in square array model and at 30° anti-clockwise from the line AD in hexagonal array model). We have found that the Tresca equivalent stresses (i.e. difference between maximum and minimum principal stresses) at the interface are maximum in magnitude at the points of maximum radial stresses.

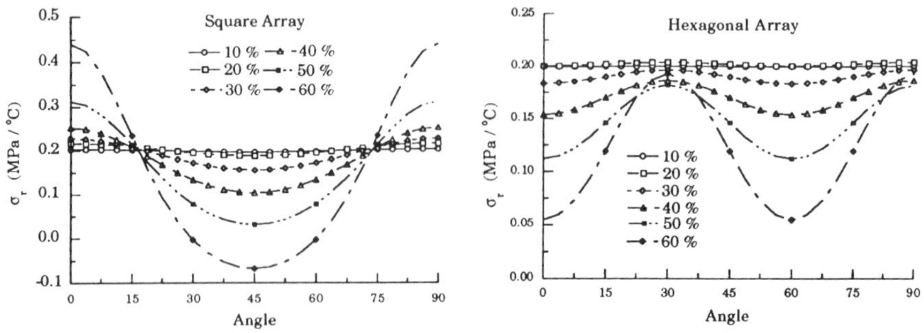

(a) Square Array (b) Hexagonal Array

Figure. 4. Residual radial stress distributions at the interface for 1° C increase in temperature vs fiber volume fraction.

For the same volume fraction, the square array arrangement of fibers results in a wider distributions of the stresses than those for the hexagonal arrangement of fibers. In fact, for higher volume fraction (> 0.50), the radial stress changes sign on some portion of the fiber-matrix interface (Fig. 4a). In the case of cooling, although one would intitutively expect a compressive stress at the interface (as the thermal expansion coefficient of the fiber is lower than that of the matrix), some portion of the interface would be under tensile stress for high fiber volume fraction.

Figure 6 shows the plot of longitudinal stress in the matrix for various fiber volume fractions for 1° C increase in temperature. For cooling, the longitudinal stress in the matrix is - $\sigma_{z\ m}$, i.e., tensile. For volume fraction < 0.4, the calculated longitudinal stress for the square array model is same as that for the hexagonal array model. For volume fraction > 0.4, the calculated longitudinal stress for the square array model is more than that for the hexagonal array model. For a given value of fiber volume fraction, the calculated longitudinal stress for both the square array and hexagonal array models is always less than that predicted by the two-cylinder model.

The magnitude of the longitudinal residual stress in the graphite / aluminum composites is in conformity with the experimental findings of Tsai et al.[2] Using stress relaxation temperature of 200° C for aluminum matrix[1] and 1.2 MPa / ^{o}C as an average value of the residual longitudinal stress (Fig. 6), the residual longitudinal stress in the graphite / aluminum composite at room temperature will be about 220 MPa which compares well with the experimental value of 200 MPa measured by Tsai et al.[2] Using 0.2 MPa / ^{o}C as an average value for the radial residual stress, the radial stress in the composite will be about 40 MPa which is lower than the experimental values of Tsai et al.[2] which ranges from 120 to 160 MPa. This difference may be due to the presence of an oxide, carbide, and / or titanium diboride layer[2] at the fiber-matrix interface. The coefficients of thermal expansion of these layers are different from those of the fiber or matrix, hence these layers change the residual stress state at the interface in the transverse direction, but their effects on the residual stress state in the longitudinal direction are negligible.

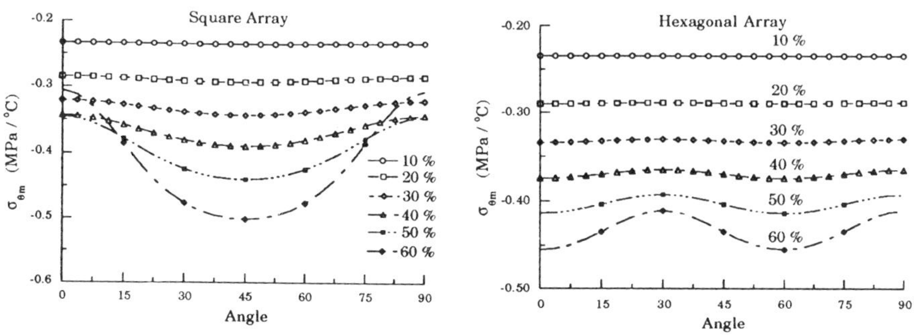

(a) Square Array (b) Hexagonal Array

Figure 5. Residual tangential or hoop stress distributions in the matrix at the interface for 1° C increase in temperature vs fiber volume fraction.

CONCLUSIONS

Finite element technique was used to analyze the residual thermal stresses in the graphite / aluminum fiber-reinforced metal-matrix composites in three dimensions. The calculated radial residual stresses were found to be significantly higher than those predicted by the two-cylinder model of Vedula et al.[1] In both the square array and hexagonal array models, the residual radial stress and Tresca equivalent stress distributions at the fiber-matrix interface have maximum magnitude at the points where the interface intersects the lines joining the fiber center to its first nearest neighbors. These interface points are possible sites for the crack initiation.

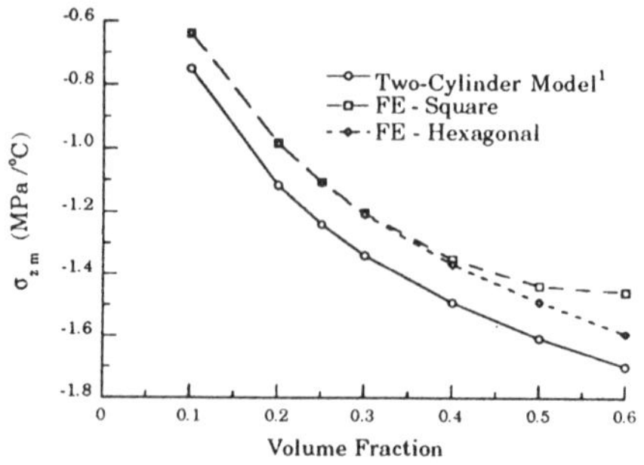

Figure 6. Residual longitudinal stress in the matrix at the interface for 1° C increase in temperature vs fiber volume fraction.

ACKNOWLEDGMENT

Computational assistance on the Cray Y-MP8/864 computer at the Ohio Super Computer Center, Columbus, Ohio, is gratefully acknowledged. This project was supported by U.S. Department of Energy, Basic Energy Sciences Division under contract No. DE-FG02-91ER45459.

REFERENCES

[1]M. Vedula, R.N. Pangborn and R.A. Queeney, Fibre anisotropic thermal expansion and residual thermal stress in a graphite / aluminum composite, Composites, Vol. 19, 55-60 (1988).

[2]S.D. Tsai, D. Mahulikar, H.L. Marcus, I.C. Noyan and J.B. Cohen, Residual stress measurements on Al-Graphite composites using X-ray diffraction, Mater. Sci. and Engng., Vol. 47, 145-149 (1981).

X-RAY TOMOGRAPHIC MICROSCOPY INVESTIGATION OF THE DUCTILE RUPTURE OF AN ALUMINUM FOIL BONDED BETWEEN SAPPHIRE BLOCKS

Wayne E. King,[1] Geoffrey H. Campbell,[1] David L. Haupt,[2] John H. Kinney,[1] Robert A. Riddle,[2] and Walter L. Wien[1]

[1] Chemistry and Materials Science Department
[2] Mechanical Engineering Department

University of California, Lawrence Livermore National Laboratory, Livermore, CA 94551-9900

(Received June 22,1995)

Introduction

Metal/ceramic interfaces play a critical role in determining the mechanical behavior of composite materials. In metal-matrix composites (MMCs), the presence of the reinforcing phase places the normally ductile matrix phase in a highly constrained state of stress. Strong interfaces between matrix and reinforcement favor crack advance by ductile rupture of the metal, while weak interfaces favor debonding. In practice, the introduction of reinforcements into a metal produces a variability in behavior that limits its commercial application (1). Because of this, micromechanics models of failure have been developed for these systems that attempt to describe the failure mechanisms.

Models for crack extension in these materials often use idealized geometries to simplify the calculations. The failure of a thin metal foil constrained between ceramic blocks has been considered in several published models, e.g., (2,3). The predictions of these models depend on the micromechanic mechanism of crack extension. For example, Varias et al. proposed four possible fracture mechanisms: (i) near-tip void growth at second phase particles or interfacial pores and coalescence with the main crack, (ii) high-triaxiality cavitation, i.e., nucleation and rapid void growth at highly stressed sites at distances of several layer thicknesses from the crack tip, (iii) interfacial debonding at the site of highest normal interfacial traction, and (iv) cleavage fracture of the ceramic. Competition among the operative mechanisms determines which path will be favored.

Ductile metals are opaque to the wavelengths of visible light. Hence, observations of the mechanisms of crack extension have not been possible in these materials. The advent of synchrotron-based X-ray tomographic microscopy (XTM) (4) has made it possible to determine the three-dimensional variations in density of specimens of several millimeters in dimension for low atomic number materials at resolutions approaching 1 μm. We have experimentally duplicated the simplified geometry of the micromechanics models and subjected the specimens to a well defined stress state in bending. The bend tests were interrupted and XTM was performed to reveal the mechanism of crack extension.

Experiment

The system of pure aluminum bonded to sapphire was chosen for its low X-ray absorption, chemical compatability, and relevance to potential MMC materials. Sapphire avoids the problems associated with grain pull-out and porosity present in polycrystalline alumina. The sapphire was purchased as cylinders of 16 mm diameter by 20 mm height with faces polished parallel to (0001) ± 1° and $\lambda/10$ ($\lambda \approx 550$ nm) flatness. The aluminum foil was purchased as 99.999% pure and 50 μm thick. The foil was laser cut into disks of 19 mm diameter. A support ring assembly made of commercially pure aluminum was attached to the perimeter of the disk by laser welding. Since bonding of the aluminum foil between two cylinders was to be accompished by ultrahigh vacuum diffusion bonding (5), the support ring facilitated the remote handling of the disks inside the diffusion bonding machine. Otherwise, the materials were used in the as-received state.

The surfaces of the sapphire to be bonded and both sides of the Al foil were sputter cleaned with 1 keV Xe^+ impinging at 15° above the horizontal. Specimens were rotated during sputtering. The surface elemental composition was characterized using Auger electron spectroscopy. All surface contaminants thus measured, including the oxide layer on the metal, were removed to below the detection limit of the technique. Previous experience with Al shows that many hours of exposure are necessary to reform a contamination layer in the UHV environment ($< 10^{-10}$ torr). The specimens were stacked and a load of 10 MPa applied. The specimens were heated to 600°C and held for 38 h. Relatively long bonding times were required due to variations in the thickness of the Al foil which requires creep to bring the surfaces into contact and eliminate all void space.

Bend beams were cut from the diffusion bonds using diamond cutting. The beams had nominal dimensions of $3 \times 3 \times 40$ mm. To ensure that failure initiated in the metal foil rather than at a flaw in the ceramic, a thin notch was placed within the metal. The notch root diameter was 40 μm and was 500 μm deep. The beams were loaded in four-point loading with inner span of 10 mm and outer span of 30 mm. Several beams were loaded to failure to characterize failure stresses.

Specimens examined with XTM were first characterized in the as-notched condition. The beam was then loaded until the first sign of non-linearity in the load versus cross-head displacement plot was observed. The mechanical test was interrupted, the specimen removed, and an XTM scan performed. The mechanical test was subsequently resumed to strain the beam further. This iterative sequence was continued for as long as was practical.

In performing the XTM scans, the X-rays from a synchrotron source (the 8-pole wiggler beamline 4-2 at the Stanford Synchrotron Radiation Laboratory) were passed through a Si(220) monochromator to select 25 keV photons. After passing through the specimen, the X-rays struck a $CdWO_4$ scintillator. The light emitted from the scintillator passed through magnifying optics and was collected by a CCD camera at a magnification sufficient to give the volume elements (voxels) in the reconstructed image the size 5.3 μm on a side. Radiographs were collected at 0.5° increments of sample rotation over a range of 180°. Reference images of the unobstructed X-ray beam were acquired every 3 degrees to correct the radiographs for beam inhomogenieties, beam instabilities, and pixel-to-pixel gain variations in the CCD array. A three-dimensional image, which represents the product of the pixel size and the absorption coefficient, was reconstructed as slices delineated by the rows of the CCD array using a filtered backprojection algorithm. Defects in the scintillator can give rise to bright or dark spots in radiographs which are not compensated for through normalization to the reference image. Such spots result in the appearance of bright or dark (usually dark) "rings" in the reconstructed slice. Most rings were removed using an interpolation algorithm on experimental sinograms. However some rings remained which can, in severe cases, give rise to contrast resembling voids, i.e., relatively low values of the absorption coefficient-thickness product. Such severe rings were only found in the scan of the as-prepared sample.

Locations of interfaces are delineated by pairs of bright and dark fringes in the XTM contrast. Although the origin of this fringe is currently not fully understood, the point of contrast reversal has been empirically shown to mark the interfaces. Voids revealed by XTM were identified and counted with an efficient, single pass cluster labeling algorithim developed by Hoshen and Kopelman (6) and described in more detail elsewhere (7).

Results

As-prepared sample

XTM revealed that the sample prepared in the 4-point bend geometry exhibited two dense interfaces between the aluminum and the sapphire. No resolvable voids were observed outside of three regions affected by ring defects. Figure 1a shows a rendering of the volume studied with XTM. The profile of the notch was extracted from this data and is plotted in Figure 2. This profile was used as input for finite element calculations to calculate the stress state of the sample.

First deformation

Figure 3a shows the load-displacement curve for the first in a series of three deformations. Loading was relaxed at the first indication of the onset of plasticity. XTM revealed four resolvable voids located at the interfaces

Fig. 1. Rendering of (a)undeformed volume, (b)after first loading, (c)after second loading, (d)after third loading. Voids are shown in color.

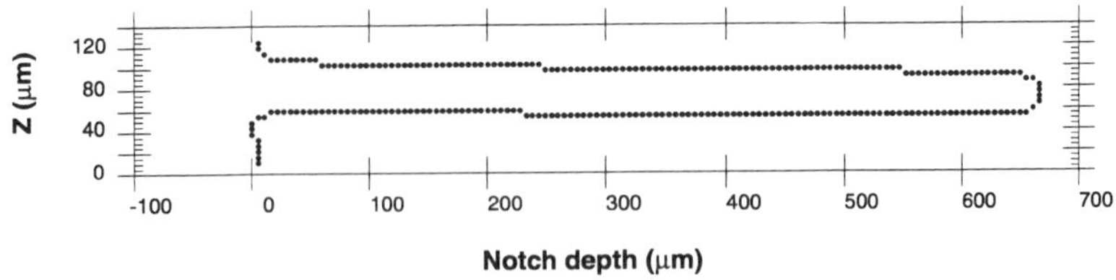

Fig. 2. Notch profile deduced from XTM.

Table I. Finite element modeling results for the maximum in triaxiality, maximum in tensile stress at the interface, and their locations relative to the notch root relative to the foil thickness.

Loading	σ_m/σ_o	X_m/h	σ_{yy}/σ_o	X_{yy}/h
1	4.1	2.3	4.5	2.3
2	5.4	2.8	5.9	2.6

Yield stress used in this calculation was 12.2 MPa, foil thickness was 50 μm, and notch depth was 630 μm.

between the metal and the sapphire. These voids were of volume between 3 and 7 voxels. Three of the voids were lenticular in nature. The fourth, and largest, had an aspect ratio (ratio of the radius of gyration normal to the interface to that in the plane of the interface) of about 1, i.e., the largest void was spherical. Figure 1b show the rendered volume with voids. Finite element modeling results for the maximum in triaxiality, maximum in tensile traction at the interface, and their locations relative to the notch root are shown in Table I.

Second deformation

Figure 3b shows the load-displacement curve for the second deformation. XTM revealed 53 resolvable voids located at the interfaces between the metal and the sapphire. As shown in Figure 4a, the distribution of void centers of gravity as a function of location perpendicular to the interfaces. Figure 4b shows that voids are concentrated at the interfaces between the metal and ceramic and are about equally distributed between the two interfaces. Voids of size less than ~30 voxels are usually lenticular in shape. Between volumes of 30-100 voxels, the aspect ratio grows to ~0.6. For volumes in excess of 100 voxels, the aspect ratio decreases to ~0.2. Figure 1c shows a

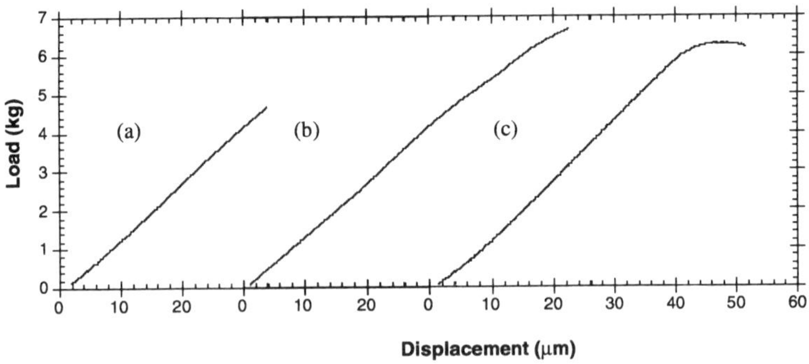

Fig. 3. Load as a function of cross-head displacement for (a) first loading, (b) second loading, (c) third loading.

(a) (b)

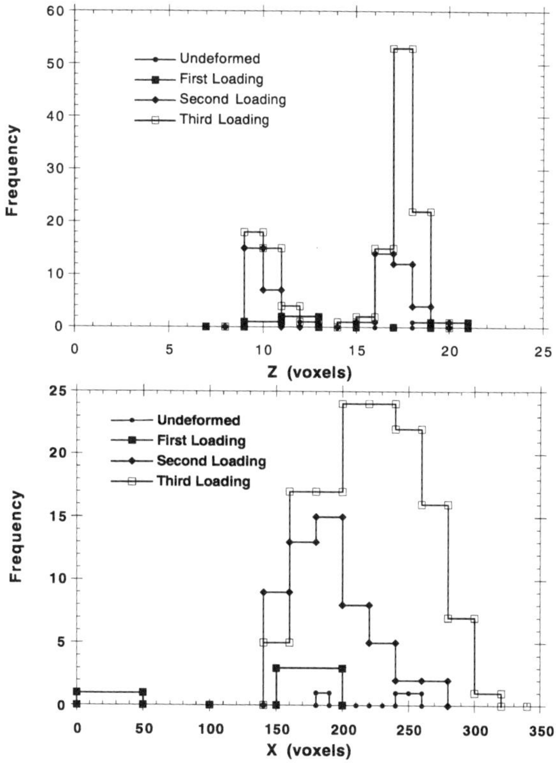

Fig. 4. Frequency distribution of voids (a) perpendicular to the interfaces and (b) parallel to the interfaces, in the direction ahead of the notch.

rendering of the void distribution and illustrates the increase in the void population compared with the first deformation.

Third deformation

Figure 3c shows the load-displacement curve for the third deformation. XTM revealed 131 resolvable voids located at the interfaces between the metal and the sapphire. Figure 4 shows the distribution of void centers of gravity as a function of location (a) perpendicular to the interfaces and (b) parallel to the interface. While some small voids do take on a spheroidal shape, the general trend for the smallest and the largest voids is to have aspect ratios of <0.2. Voids in the range of 200-900 voxels in volume tend to have aspect ratios approaching 0.4. The void population has dramatically increased and the voids have spread ahead of the notch (Figure 1d).

Discussion

XTM has been used to elucidate the mechanism of ductile rupture of a 50 μm thick Al foil bonded between two sapphire cylinders subjected to Mode I loading. At the onset of plasticity, small lenticular voids form at each interface at a location consistent with the maximums in triaxiality and tensile traction at the interface. As such, this work reveals that debonding plays an important role in the failure at the earliest stages of plasticity. Previous observations of rupture in this system revealed the presence of voiding on the fracture surfaces, e.g. see Ref. 8 and 9.

However, the authors were only able to link the origin of some voids with the interfaces, i.e., the generalization of this observation to all voids was not clear. In this investigation, all voids were initially associated with an interface. Further, using conventional fractography, the mechanism of void nucleation and growth could not be revealed. The lenticular nature of these voids as revealed by XTM suggests that they may be formed by debonding and propagation of an interface crack rather than by nucleation and growth via dislocation agglomeration. The existence of debonds at interfaces in Al is consistent with observations in other systems, e.g., Au (10), Ni (11), and Cu (12), which debond but do not exhibit the extensive ductile dimpling found in Al. The observations of this work suggest that the tendency of a system to rupture in a ductile or brittle fashion may be controlled by the interplay of the interface bond strength and the yield strength of the metal.

As deformation proceeds, voids of intermediate size tend to become spherical under the influence of the triaxial stress state which serves to relieve the constraint locally. It is reasonable to conclude that the maximum in the constraint then shifts to locations farther in advance of the notch root where more small debonds form. At high strains, spherical voids continue to grow until they interpenetrate which eventually leads to the formation of the ductile ligaments observed in fractographs. There appears to be a second class of voids that remain lenticular throughout the deformation. These seem to be lenticular voids that have grown past some perhaps critical size above which the tendency to become spherical is diminished.

Conclusions

Ultrahigh vacuum diffusion bonding has been used to make model specimens in the Al/sapphire system. The geometry of a thin, ductile foil sandwiched between strong, elastic ceramic blocks is identical to that modeled by several investigators (2,3). The miocromechanics of failure for this system were observed by XTM. The primary findings are that (i) damage ahead of the notch initiates by interface debonding at a location coinciding with maxima in triaxiality and tensile traction at the interface, (ii) debonding occurs at the most early stages in the observation of plasticity, (iii) the debond expands for a limited distance, likely arresting due to crack tip blunting, (iv) this lenticular debond then becomes spherical with further strain and (v) intergrowth of the spherical voids leads to the typical ductile rupture fracture surfaces observed in this system. Future efforts to model this effect will need to include the debonds at the interface which alter the state of stress.

Acknowledgements

The authors grategully acknowledge Professor R. H. Dauskardt for providing the mechanical testing equipment used in this investigation. This work performed under the auspices of U. S. Department of Energy and the Lawrence Livermore National Laboratory under contract No. W-7405-Eng-48.

References

1. W. E. King, et al., Mater. Sci. Eng. A191, 1 (1995).
2. A. G. Varius, Z. Suo, and C. F. Shih, J. Mech. Phys. Solids 39, 963 (1991).
3. V. Tvergaard and J. W. Hutchinson, Phil. Mag. A 70, 641 (1994).
4. J. H. Kinney and M. C. Nichols, Ann. Rev. Mater. Sci. 22, 121 (1992).
5. W. E. King, et al., Mat. Res. Soc. Symp. Proc. 314, 61 (1993).
6. J. Hoshen and R. Kopelman, Phys. Rev. B,15 3438 (1976).
7. J. H. Kinney, N. E. Lane, and D. L. Haupt, J Bone and Miner Res. 10, 254 (1995).
8. B. J. Dalgleish, K. P. Trumble, and A. G. Evans, Acta Metall. 37, 1923 (1989).
9. W. E. King, G. H. Campbell, S. L. Stoner, and W. L. Wien, Cer. Eng. Sci. Proc., 15, 769, (1994).
10. I.E. Reimanis, B.J. Dalgleish, and A.G. Evans, Acta Metall. Mater. 39, 3133 (1991).
11. I.E. Reimanis, Scripta Metall. Mater. 27, 1729 (1992).
12. T.S. Oh, J. Rödel, R.M. Cannon, and R.O. Ritchie, Acta Metall. 36, 2083 (1988).

KEYWORD AND AUTHOR INDEX